DISCOVERING
THE BRAIN

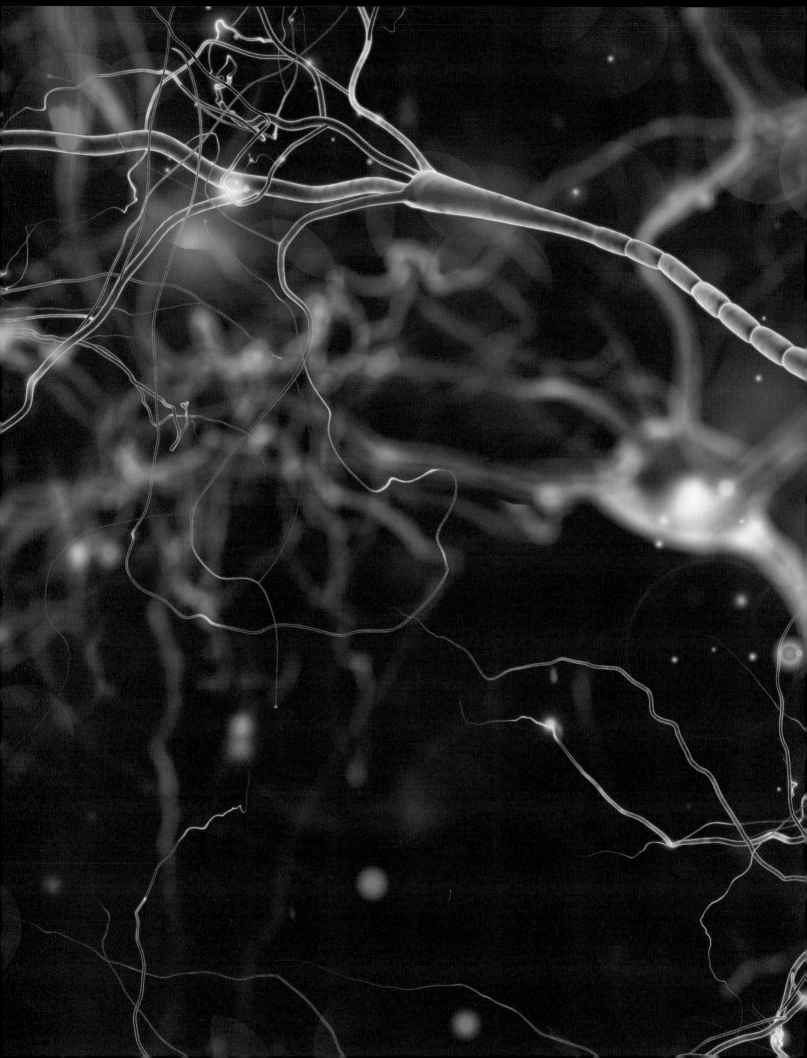

DISCOVERING THE BRAIN

A guide to the most complex organ of the human body

FRANK AMTHOR

SIRIUS

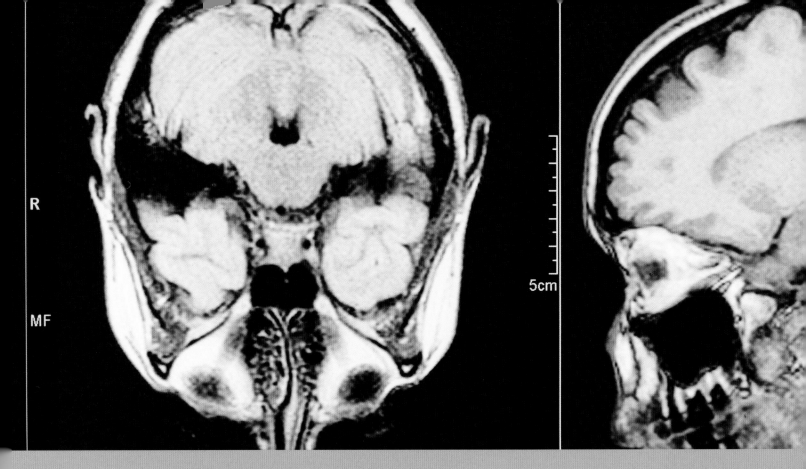

5cm

R

MF

SIRIUS

This edition published in 2023 by Sirius Publishing, a division of
Arcturus Publishing Limited,
26/27 Bickels Yard, 151–153 Bermondsey Street,
London SE1 3HA

ISBN: 978-1-3988-2075-3
AD008835US

Printed in China

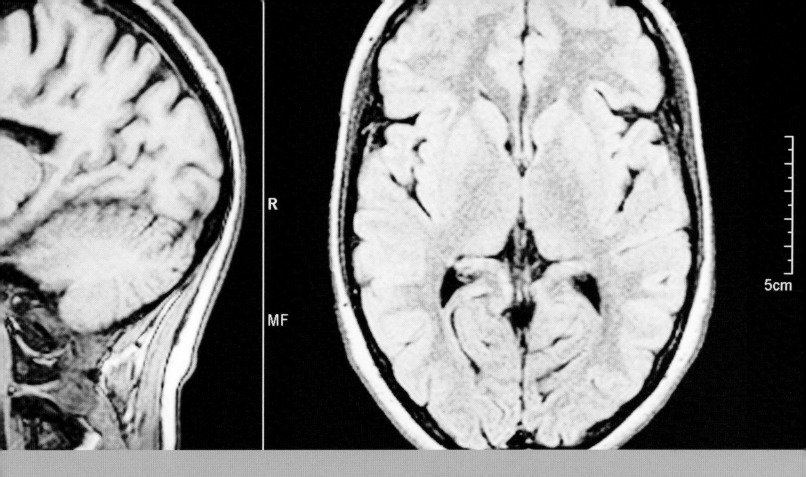

R
MF
5cm

Contents

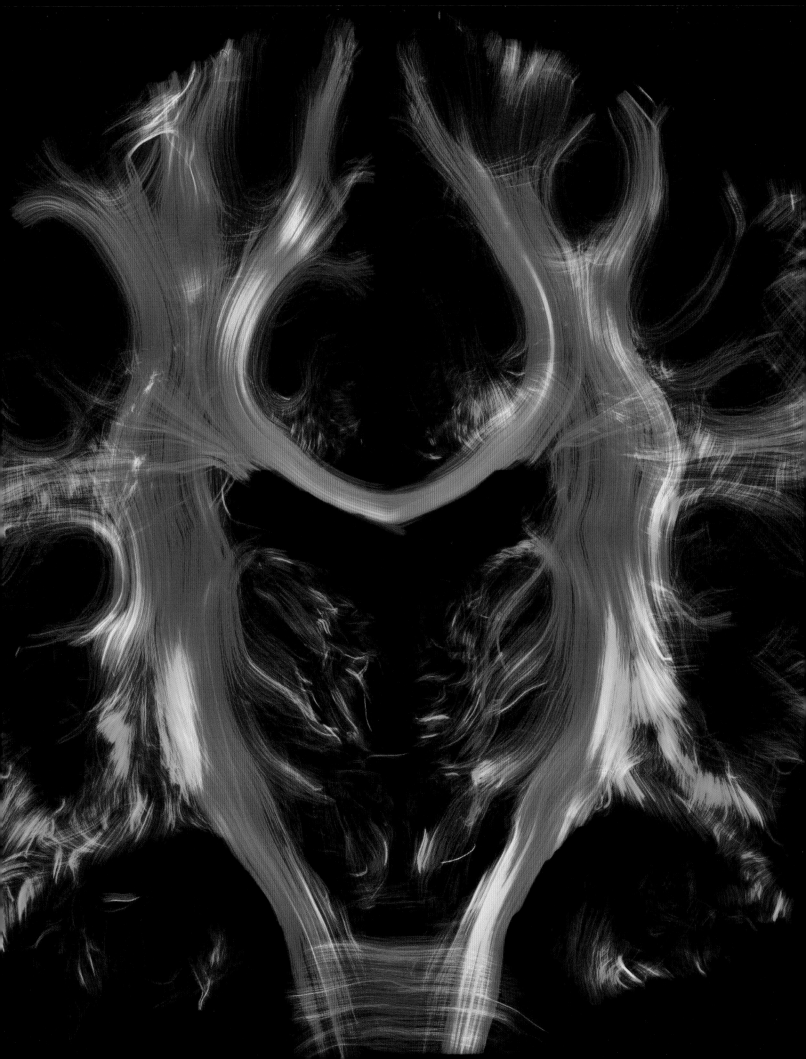

// Introduction

What does it mean, to discover the brain? The first part of the answer has to do with understanding what the brain does. Aristotle in the 4th century BCE thought that the heart was the seat of intelligence, but it was clear by the Middle Ages that the heart is a pump. There was in ancient times, however, no simple mechanical analog for the brain.

When steam and pneumatic systems were developed in the late Middle Ages, axon tracts in the brain were postulated to be little pneumatic tubes that transferred sensory input or motor commands within the body like the tubing around a steam engine. Descartes supposed that the optic nerve communicated vibrations corresponding to the visual image from the eye to the brain, where the pineal gland, which housed the soul, viewed the images, leading to perception.

A central idea associated with Descartes' treatise is the concept that the human brain is a machine, like the brain of any animal, but for humans the brain is controlled by an immaterial soul, which animals lack. The history of neuroscience has more or less followed the trajectory of assigning more and more function to the animal machine brain, and less to the soul. The materialist view at the extreme of this process is that there is no soul and that humans differ from animals only quantitatively, not qualitatively.

Most neuroscientists don't really take a position on whether there is a soul or not, but rather, try to see how much a wetware machine can explain about human behavior without requiring a soul in the pineal gland. This idea is that the brain is a kind of wetware analog computer that receives and processes information via axon tracts that send electrical pulses, instead of pneumatic pressure, throughout the body.

The questions that follow this idea, then are:

1. Can we simulate animal and human brains by computers that simulate the analog processes that occur in brains?
2. What do we need to know about the brain to do such a simulation?
3. Will good simulations of the brain allow us to understand how it works, repair damaged brains, build better artificial brains, or enhance human intelligence?

In this book, we will examine how scientists have answered these questions and created our modern understanding of the brain. The fundamental computing unit of the brain appears to be the synapse, thousands of which "operate" each neuron. Millions of neurons form groups called nuclei or ganglion that process inputs, perform calculations of some kind, generate motor behavior, and construct memories, awareness, and consciousness.

Below: *Descartes' conception of the brain as illustrated in his* Treatise on Man, *written in the 1630s and published posthumously in 1664.*

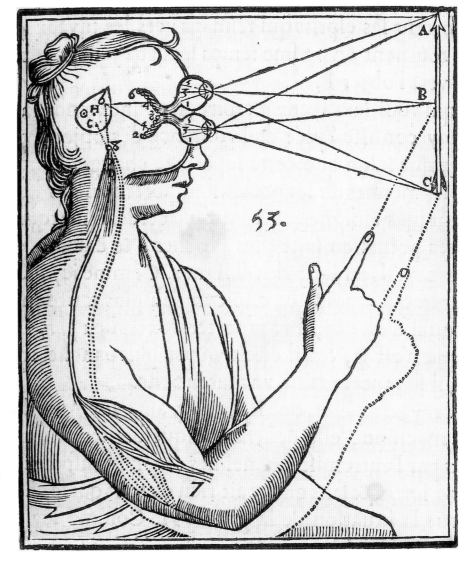

A BRIEF HISTORY OF BRAIN SCIENCE

The mysteries of the brain have fascinated men and women across the globe for millennia. As far back as the Neolithic era we can find evidence of early attempts at brain surgery, while leading classical scholars, such as Hippocrates and Galen, sought to deepen our understanding of the nature of human thought. In the 19th century more systematic analysis of the biology of the brain by pioneers like Korbinian Brodmann, Paul Broca, and Karl Wernicke laid the foundations for our modern understandings of neuroscience. But it was not until the 20th century that we began to truly develop the tools and techniques that allow us to understand the inner workings of the brain.

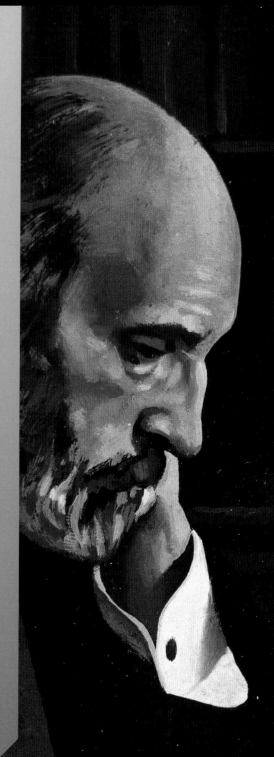

Santiago Ramón y Cajal, the father of modern neuroscience who revealed the secrets of the neuron.

// Early ideas about the brain

Little is known about what the ancients thought about the brain prior to the invention of writing. However, ancient graves yield evidence that ancient brain surgery was practiced. By far the most common ancient brain surgery was trepanation, the scraping or drilling of a hole in the skull. It is likely that this surgery was done to relieve pressure felt by the patient in his head, such as might occur from a blow to the head. Trepanation may also have been done for the purpose of "expelling evil spirits," which were thought to be the cause of migraine headaches, epilepsy, or mental illness.

Neolithic trepanation

Ancient skulls with unmistakable evidence of trepanation, and recovery afterward, have been found in both the old and new world.

Below: *A trepanated neolithic skull, c. 3500 BCE.*

Trepanned skulls have been found in Europe from 5,000–7,000 years ago. Forensic examination shows that the trepanation was done with a rotating or scraping tool, such as a flint knife, which can be sharper than modern scalpels. Healing afterward shows that many of these patients recovered from the surgery. The surgical holes were covered with new bone, and there was, in some cases, little sign of infection. The healing process in these specimens must have been on the order of several years. The lack of infection is quite remarkable for such ancient surgery.

Written records about the brain

Among ancient western civilizations, the earliest known written references to the brain are from Egypt. A 17th century BCE Egyptian papyrus refers to battlefield head wounds and their effects. The ancient Egyptians also used trepanning with a variety of techniques, ranging from flint knives to

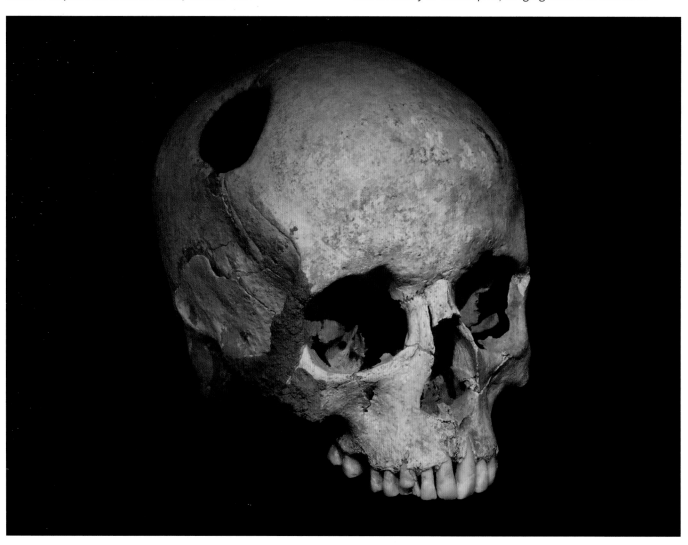

Above: *The Edwin Smith papyrus from ancient Egypt, dating back to c. 1600 BCE, showing the first use of the word brain in writing.*

a specialized trepanning tool consisting of a hollow tube with serrated edges on the bottom that was rotated against the skull. This tool was adopted later by ancient Greek and Roman surgeons. Later, in the middle ages, a central spike was added to the tool to better center the rotary movement.

Although Aristotle in the 4th century BCE thought that the heart was the seat of intelligence, earlier writers, such as Alcmaeon, had postulated that intelligence and sensory perception depended on the brain. Two notable Greeks in the history of brain science were Hippocrates and Galen.

Hippocrates (460–377 BCE) lived during the pinnacle of classical Greek civilization. He differentiated, by clinical observation, medicine from mysticism and theology. His treatise "On Injuries of the Head" systematically discussed cranial anatomy and results of head trauma, especially from military operations. He recommended strategies for surgical management of head trauma including trepanation, incisions, and wound treatments.

About six centuries later, Galen (129–216 CE) studied anatomy and physiology at the Alexandrian school of medicine, after which he became a physician charged with treating wounds of gladiators. From this and other experiences he authored over 500 books and treatises. His writings describe the phenomenon of hydrocephalus (a build-up of fluid in the brain), the pineal and pituitary glands, tectum, corpus callosum, fornix and ventricular system. He delineated the central from peripheral nervous systems (basically, nerves inside versus outside the brain and spinal column), and the fact that spinal trauma can cause limb paralysis. He was clear about the idea that mental activity occurred in the brain, not the heart, as Aristotle had suggested.

Middle Ages and Renaissance

A dominant idea about brain function in the European Middle Ages was that the brain was organized around three ventricles, each responsible for a different aspect of mental function. Imagination was mediated by the anterior ventricle and memory in the posterior ventricle. Reason and common sense was located somewhere in the middle of these.

Knowledge of the anatomy of the brain exploded in the Renaissance when physicians began routinely to do post-mortem brain dissections. Leonardo da Vinci made meticulous drawings of his own brain dissections and through them began to suspect that the ventricle theory was lacking.

A leap in brain science occurred when the English physician Thomas Willis published his *Cerebri Anatome (Anatomy of the Brain)* in 1664. This work can be considered one of the modern foundations of clinical neuroscience. Many of the terms Willis used for different brain structures continue to be used today.

Willis made a huge contribution to understanding brain function by showing details of brain vasculature and circulation visualized by the use of India ink injections, as William Harvey had done for blood circulation in other parts of the body.

Below: *Hippocrates and Galen.*

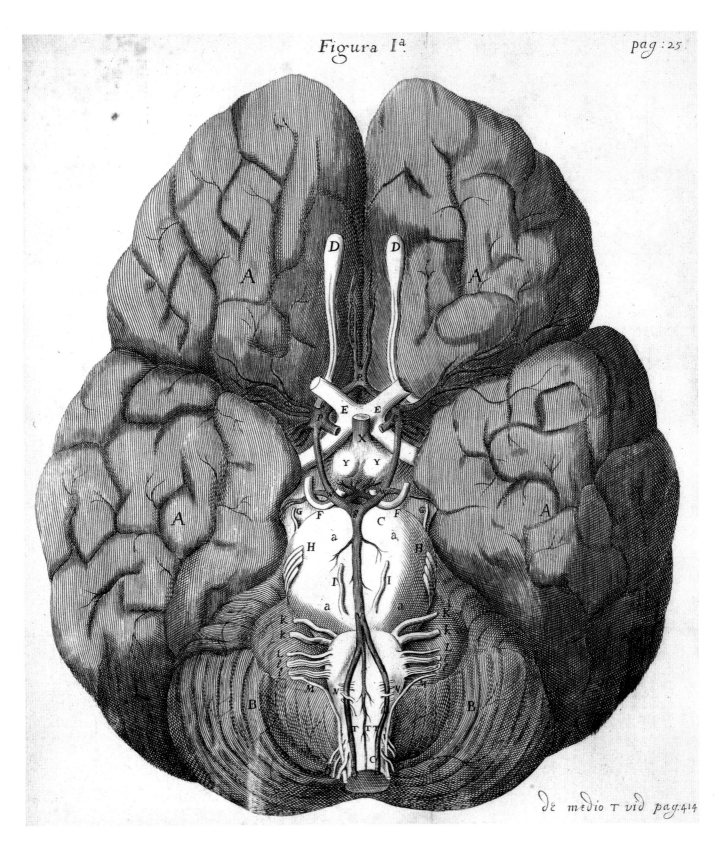

Above: *A Willis drawing of the ventral side of the brain done by Christopher Wren, the famous architect, showing details of the optic and olfactory nerves and other ventral brain structures, in previously unknown accuracy.*

// The localization of brain function

Ancient studies of the brain and the effects of injury on the brain produced three fundamental conclusions:

1. The brain has different areas.
2. Different brain areas differ in physical appearance.
3. Different areas of the brain, when damaged, have different effects.

Since the time of Galen it was clear that there were differences between the brain and peripheral nervous system, although the two were functionally connected. Moreover, within the brain, there were demonstrably different structures such as the brainstem and cerebellum, thalamus, basal ganglia, and cerebral cortex (see pages 48–55).

The functional anatomy of the cerebral cortex remained a mystery until the late 19th century for several reasons:

1. Non-mammalian vertebrates had virtually no cerebral cortex, and, although apparently less intelligent than mammals, clearly exhibited complex perceptual and behavioral capabilities.

2. Within mammals there was a large variation in the size of the cerebral cortex. Mammals thought to be phylogenetically old, and therefore less intelligent, had smaller cortices than more advanced mammals, especially primates.

3. Within primates there appeared to be a steady increase in cortex size from monkeys to apes to humans. If humans are the most intelligent of mammals and have the largest cerebral cortex, there must be some dependence of intelligence on the cortex.

A major problem in understanding the role of the cerebral cortex was that it was very large and not strongly differentiated in appearance from area to area. Two major theories arose about cortical function, called *localization and associationism*.

Localization: subdividing the cortex

In the 1800s, particularly in Germany, a number of techniques for staining biological tissue were developed, many of which were derived from newly invented textile dyes.

Below: *The Brodmann areas of the brain.*

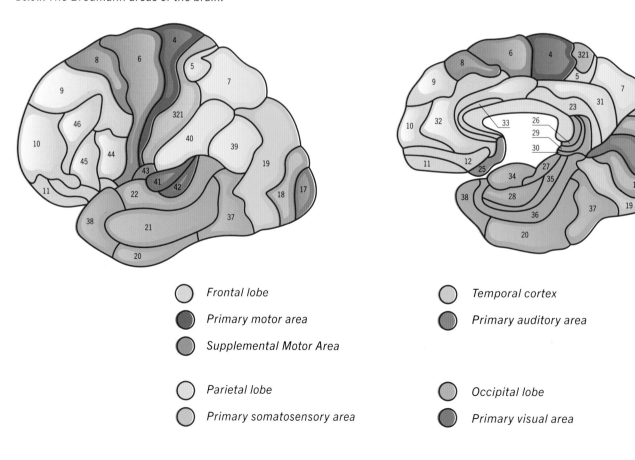

○ *Frontal lobe*

● *Primary motor area*

◐ *Supplemental Motor Area*

○ *Parietal lobe*

○ *Primary somatosensory area*

○ *Temporal cortex*

◐ *Primary auditory area*

○ *Occipital lobe*

● *Primary visual area*

The study of tissue via staining is called histology. When applied to the brain, some of these new dyes stained neural cell bodies only, while others stained only axons (the portion of a neuron that carries signals from the cell body to the axon terminal, where synapses occur). Use of these dyes permitted differentiation between nuclei or ganglion, on one hand, and axon tracts on the other.

The cell body dyes were particularly useful because they showed that the organization of cell bodies was different in different cortical areas due to differences in soma size, shape, and layering of cells. Some of these differences were found to be consistent in particular brain areas for animals of the same species, and even consistent for these brain areas between related species. The differentiation of brain areas using cell body dyes is called *cytoarchitectonics*.

Korbinian Brodmann (1868–1918) was one of the many researchers who studied brain cytoarchitecture with various dyes. He divided the cerebral cortex into 52 regions, many of which were homologous across different mammalian species. Each of the four major lobes of the cortex (frontal, parietal, temporal, and occipital) (see pages 54–5) is comprised of a distinct set of Brodmann areas. The Brodmann numbering scheme has no particular significance in its order because Brodmann numbered the areas in the order in which he studied their cytoarchitectonics. However, nearby areas are typically numbered consequently.

Brodmann areas 1–3 constitute the somatosensory area of the most anterior portion of the parietal lobe, which is just posterior to the central sulcus. Area 4, on the other hand, which is the primary motor cortex, is the most posterior portion of the frontal lobe.

A fundamental idea of the Brodmann (and other) numbering scheme is that:

1. Areas that have a very different cytoarchitectonic appearance are likely to have different functions.

Above: *Paul Broca.*

Above: *Carl Wernicke.*

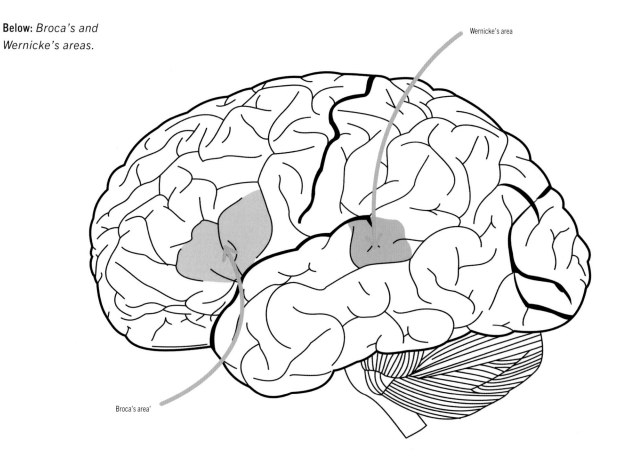

Wernicke's area

Broca's area'

2. Contiguous areas with similar cytoarchitectonic appearance are likely to have a similar function, but displaced across the body, or visual space, or auditory frequency, or along whatever parameter that cortical area is involved in processing.

The Brodmann numbering scheme allowed neurologists and neuroscientists to be much more specific about discussions of brain area and function. This soon lead to the discovery of function localization to specific brain areas, and ultimately, to the extreme view of *phrenology*, where very subtle personality attributes were thought to be localized to specific cortical areas. One of the first functions to be localized to a specific brain area was that for language production, demonstrated by Paul Broca (1824–80), a French physician and anatomist and Carl Wernicke (1848–1905) (see pages 168–9).

Phrenology

Given Brodmann's anatomical division of the cortex into 52 distinct regions, the question arose as to whether each had a specific function. Franz Joseph Gall (1758–1828), an Austrian physician and neuroanatomist, postulated that the brain was organized around some 35 specific

functions. These functions included not only cognitive capabilities such as language and color perception, but in addition, more ephemeral capacities such as hope and self-esteem. The general term for this idea is phrenology.

Gall and colleagues, such as Johann Spurzheim (1776–1832), also thought that if a person frequently used some mental faculty, the part of the brain mediating that function would grow, and that the expansion of this brain area would then cause a bump in the overlying skull. Thus, a careful analysis of the skull could describe that person's personality, calling this technique anatomical personology. Essentially, the theory of phrenology is that learning is translated into brain growth that pushes the skull outward.

Unfortunately, there was little evidence for localization of functions like sublimity or cautiousness to the assigned brain regions. Many of these assignments were taken from single cases of brain injury, or, even worse, from relationships in single individuals between low or high function in some particular trait and their skull shape.

The excesses of phrenology contributed to another set of theories that were excessive in the opposite direction. These are called *associationism*.

Opposite: *Spurzheim's phrenological map.*

13 Firmness   14 Reverence   15 Benevolence
16   17   18 Marvellousness   21 Imitation
Hope   Conscientious-ness
12 Self Esteem
11 Approbativeness
10   8   19 Ideality   35   Causality   22 Comparison
Cautiousness   Acquisitiveness   20 Mirthfulness
7   27 Locality
Secretiveness   32   31   Eventuality
3   6   Tune   Time
Philoprogenitiveness   Adhesiveness   Constructiveness   24   Individuality
5   Combativeness   4   9   28   23   25   34
Destructiveness   29   Order   Colouring   Size   Weight
2   Aliment-   Calculation   33 Configuration
Amativeness   iveness
1 Inhabitiveness   30 Language

Pendleton's Lithog'y Boston.

DR. SPURZHEIM.
Divisions of the Organs of Phrenology marked externally.

Entered according to Act of Congress in the Year 1834, by Wm S Pendleton in the Clerks Office of the District Court of Massachusetts.

Associationism

Although Aristotle was wrong about the seat of intelligence being in the heart, rather than the brain, he founded a lasting school of thought about intelligence called empiricism that contrasted sharply with the idealism of his teacher Plato. Aristotle's empiricism is the idea that all knowledge comes from sensory experience that produces simple ideas and concepts. Further, when simple ideas interact and become associated with each other, complex ideas and concepts are created in an individual's knowledge system. The empiricist idea of the essential role of experience continued through the ideas of the British philosophers from Thomas Hobbes in the 17th century, through John Locke and David Hume to John Stuart Mill in the 19th century. The associationist/behavioral school of experimental psychology arose from empiricist philosophy.

The empiricists, and later associationists, argued that the brain was a general purpose learning machine whose

Below: *Karl Lashley.*

Above: *Hermann Ebbinghaus.*

capacities developed from experience. They thought that, because different people have different experiences and learn different things, there was no reason to suppose that any particular ability or knowledge should be localized to any brain area. They thought that the brain operated in a holistic manner, with functions acquired by learning stored throughout the cortex. This view was reinforced by the experiments of Karl Lashley (1890–1958), who examined the retention of maze learning in rats (see pages 156–7).

Associationism became a dominant paradigm for explaining behavior, particularly in American psychology departments. If, as the associationists postulated, cognitive, memory, and perceptual capabilities are not localized to specific areas of the brain, these capabilities should be studied and understood in terms of the relationship between stimulus sequences and contingent rewards and punishment. This was the underlying theory behind behaviorism, as studied using conditioning experiments (see pages 149–50).

Even as early as the late 1800s Hermann Ebbinghaus (1850–1909) argued that complex processes like memory could be understood and analyzed through behavioral reward contingencies. Edward Thorndike's classic 1911 monograph *Animal Intelligence: An Experimental Study of the Associative Processes in Animals* gave evidence for his law of effect, a general postulate about the nature of association and learning. Behaviorism hypothesizes that a response followed by a reward would be ingrained in the organism as a habitual response. However, if there was no reward following a response, the response would disappear. Rewards provided the mechanism for establishing adaptive responses, in a manner somewhat like Charles Darwin's natural selection was the mechanism for evolution.

// The discovery of neurons and neural activity

Luigi Galvani (1737–98) was an Italian physician and scientist who discovered that the leg muscles of a dead frog would twitch when touched by certain metal cables. He called the phenomenon "animal electricity," believing that the metal wires stimulated an intrinsic electrical potential for muscles to contract. In this he was only partly right.

Alessandro Volta (1745–1827) repeated the experiments of Galvani but came to a different, and in some ways, more important conclusion. Volta had been independently working on combinations of metals in electrolytes that acted as batteries. He concluded that the metal cables Galvani used to stimulate the frog legs acted as a battery that generated an electrical current that caused the twitching. In this he was correct. However, it is also the case that the freshly dead muscles had the ability to amplify a small electrical current to produce the muscle action potential that caused the twitch.

Taking the work of Galvani and Volta together are two significant scientific discoveries:

1. Dissimilar metals in an electrolyte (weak acid) comprise a battery that can produce electrical current.
2. Animal muscles (and nerves) can be activated by weak electrical currents to produce action potentials that greatly amplify the effects the weak current alone would produce.

Below: *Galvani's experiment with frog legs.*

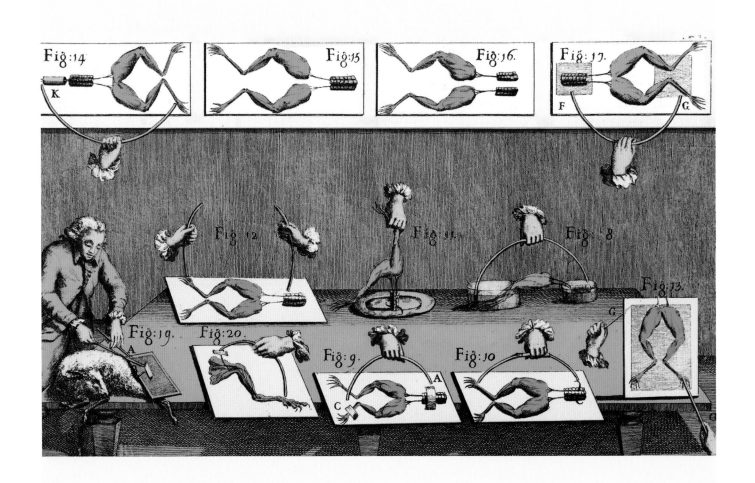

Brains are made of neurons: Golgi and Cajal

Although enormous strides were made in delineating gross brain anatomy with stains developed in the 19th century, there was a major problem with these stains when it came to describing neurons at the microscopic level. Most stains were useful for seeing neural somas, but did not reveal neural dendrites or axons. Those stains that did bind to structures in dendrites stained all the dendrites of all neurons in any given area, showing up as a black, undifferentiated mass under the microscope.

This problem was solved by the discovery of the reduced silver stain by Camillo Golgi (1843–1926). For reasons that are still not entirely clear, the Golgi stain, as it is now called, stains a small percentage of neurons in any given tissue section completely, that is, the soma, dendrites, and usually at least a portion of the axon. Because the rest of the tissue is clear, detailed drawings can be made of the dendritic architecture of individual neurons. Unfortunately, Golgi himself did not understand this. He thought the stain he had invented was just staining a portion of a continuous network (syncytium) of neural tissue.

The breakthrough in understanding single neuron anatomy was made by Santiago Ramon y Cajal (1852–1934). Because he appreciated that the Golgi stain was revealing the complete dendritic architecture of single neurons, he studied and classified the various dendritic morphologies that neurons exhibited throughout the brains of numerous animals. Cajal's detailed and accurate drawings are still used by neuroanatomists today. Cajal also realized that neurons have a functional polarity: they receive inputs on the dendrites, and send output via the axon.

The fundamental difference in understanding dendritic morphology revealed by the Golgi stain led Golgi and Cajal to be professional enemies. Although the two jointly received the Nobel Prize in physiology or medicine in 1906, Golgi refused to speak to Cajal.

Right: *Golgi's stained neurons, from a dog's olfactory bulb.*

// Modern techniques for brain study

Modern techniques for studying the brain include technologies that greatly enhance neuroscientists' ability to observe both "function" of the brain and its "structure." These techniques exist for orders of magnitude scales for both time and space.

FUNCTION

MICROELECTRODE RECORDING
Microelectrode recordings can be either extra- or intracellular. Extracellular recordings pick up the currents generated by action potentials with microelectrodes whose tips are placed near the neural soma. These recordings have sub-millisecond resolution, and, in the case of chronically implanted electrodes, can record for months. There are two major types of intracellular microelectrodes, called *"sharp"* and *"patch"* recording respectively. *Sharp recording microelectrodes (micropipettes)* have tips smaller than one micrometer in diameter, and are literally pushed through one side of the outer membrane of the neuron until the tip is inside, where the potentials generated by the neuron can be recorded. Patch pipettes have tip diameters of a few microns and are pushed against the neural membrane to record currents through it.

OPTICAL IMAGING
Optical imaging can be done to examine activity in large brain areas. It requires exposing the brain, however, and is mostly used in animal research

INTRINSIC IMAGING
High-resolution microscopic cameras can, under some circumstances, detect subtle changes in neural morphology optically, often using infrared light. Detectable types of changes include swelling during activity, and movement of portions of the neuron during activity.

DYE-BASED FUNCTIONAL IMAGING
The introduction of dyes in the neural cytoplasm or membrane can allow the optical monitoring of neural activity in real time. Almost all such dye-mediated recordings are based on changes in dye fluorescence, the re-emission of light after absorption from a stimulating light source. There are two major types of dye-mediated physiological recording: (1) ion concentration dependent changes in dye fluorescence, and (2) membrane potential induced changes in dye fluorescence.

ION CONCENTRATION FLUORESCENT DYES
Fluorescent dyes can detect changes in ion concentration. When neurons fire action potentials there is movement of sodium, potassium, and calcium across the cell membrane that can be detected by such dyes.

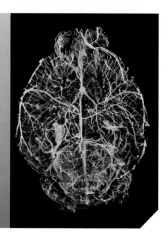

GENETIC METHODS
Genetic techniques allow the insertion into the DNA of neurons codes for protein complexes that fluoresce in response to voltage changes in the neuron or ion concentration changes. This technique is especially powerful when the genetic coding is such that the fluorescent proteins are expressed only by a specific neural cell type in a particular tissue. DNA can also be similarly inserted that makes specific neurons express ion channels whose opening is modulated by light.

FMRI

The most widely used technique for monitoring brain activity is functional Magnetic Resonance Imaging, or fMRI. fMRI is based on the paramagnetic properties of hemoglobin in the blood. Early fMRI imaging detected the amount of hemoglobin in various parts of the brain during cognitive or perceptual tasks, under the hypothesis that blood supply would increase to metabolically active areas. A newer method, called Blood Oxygen Level Dependent (BOLD) fMRI detects the ratio of oxygenated to non-oxygenated hemoglobin. High metabolic activity transfers oxygen from the blood to brain tissue, after which there follows an increase in blood flow, so the BOLD signal has better time resolution than the blood flow signal.

PET, SPECT AND SIMILAR

These techniques usually detect a correlate of physiological activity, such as glucose or oxygen consumption. The typical approach is to take a baseline activity picture, and then a second "snapshot" of activity during the performance of some cognitive or perceptual task. These techniques have the advantage of being direct measures of activity, and often have high spatial resolution. However, they have little temporal resolution as the image consists of two snapshots, before and after.

STRUCTURE

EEG

Electroencephalography was invented before the 20th century. Large areas of the brain have enough synchronous activity to produce a detectable signal on the scalp that can be picked up by flat electrodes there about the size of a coin. Because the electrical current produced by any localized source within the brain is diffused by the rest of the brain and the scalp, EEG has poor spatial resolution. The temporal resolution, however, can approach milliseconds. A derivative of the EEG, the ERP, is an averaged potential locked to some stimulus that removes some of the ongoing intrinsic signal generated by the brain to obtain a signal specifically associated with the stimulus.

X-RAYS AND CAT SCANS

Prior to the 20th century, knowledge of brain activity was derived mostly from post-mortem examinations. With the invention of X-ray detectors it became possible to image living brains, although with poor spatial resolution. Much better spatial resolution was obtained with the development of Computer Axial Tomography (CAT) scanning that allows differentiation of structures within the brain.

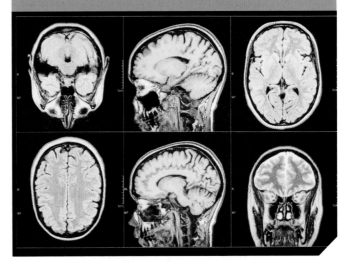

MEG

Magnetoencephalography detects the magnetic fields generated by currents within the brain. MEG signals are like EEG signals except they are much better localized because the brain and scalp do not diffuse the magnetic field signal. MEG requires expensive detectors called SQUIDs and special electrically and magnetically shielded rooms.

MRI

Magnetic Resonance Imaging now gives superior structural resolution even compared to CAT scanning. When fMRI is done, the sequence is that a high resolution MRI scan is done first, then the lower resolution functional fMRI scan is superimposed on the MRI. Computer processing can generate either a sagittal or a coronal section.

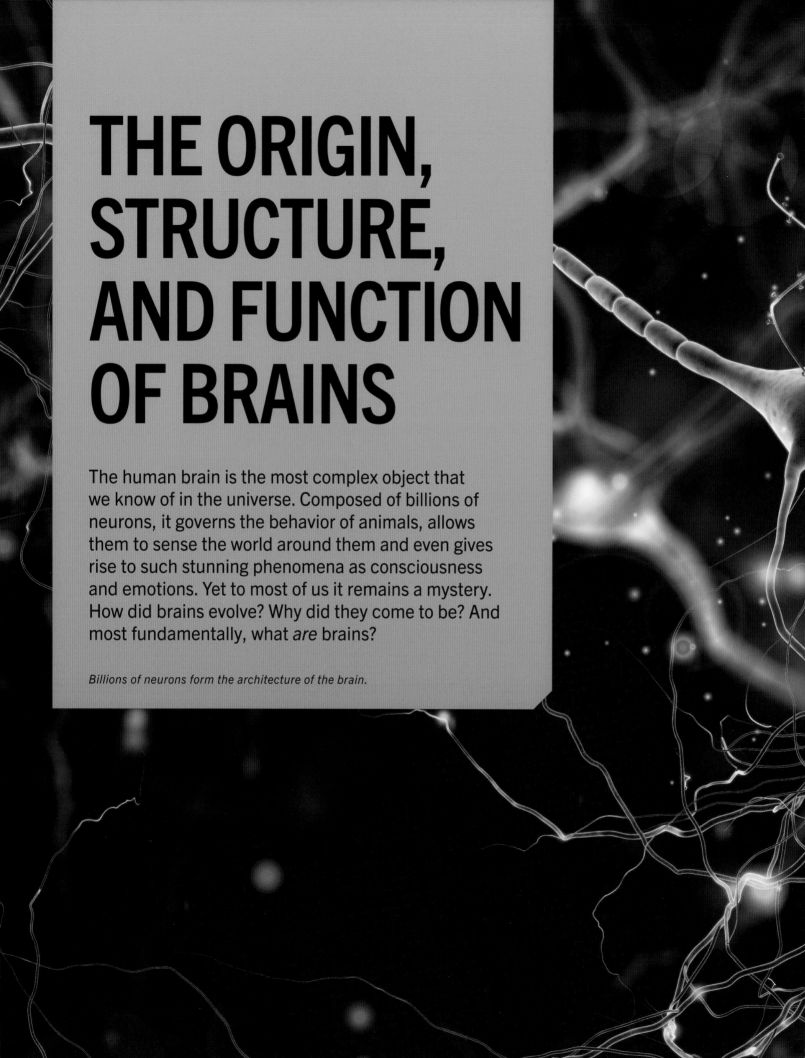

THE ORIGIN, STRUCTURE, AND FUNCTION OF BRAINS

The human brain is the most complex object that we know of in the universe. Composed of billions of neurons, it governs the behavior of animals, allows them to sense the world around them and even gives rise to such stunning phenomena as consciousness and emotions. Yet to most of us it remains a mystery. How did brains evolve? Why did they come to be? And most fundamentally, what *are* brains?

Billions of neurons form the architecture of the brain.

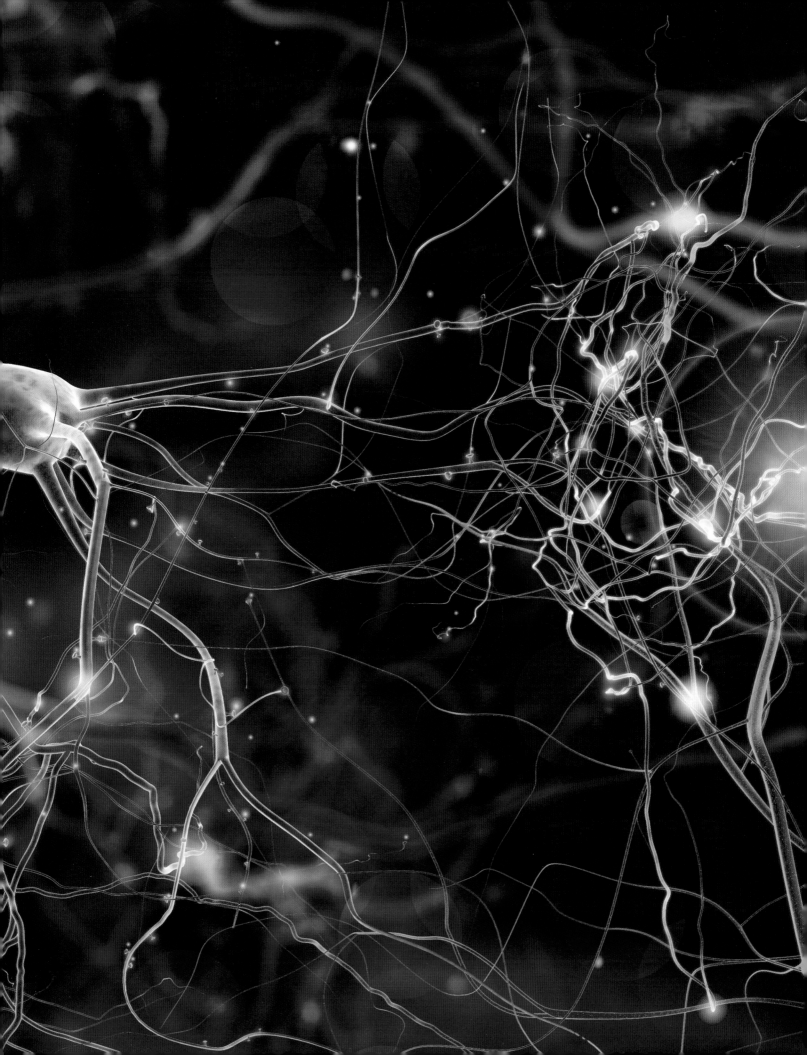

// Early life on Earth

When the earth first formed, its surface was a hot magma where life was impossible. It took nearly 750 million years before the first single-celled organisms called prokaryotes emerged. About 1.75 billion years before present, a major advance in the complexity of life occurred when eukaryotes arose. Eukaryotes are cells with nuclei that sequester their DNA from the rest of the cell. They also have other specialized complex organelles within the cell, such as mitochondria that generate adenosine triphosphate (ATP) for cellular energy needs. For the next billion years, life on earth consisted of only the single-celled prokaryote and eukaryote forms. Stimulus-response mechanisms in single-celled organisms are primarily biochemical. Toxins or food sources either bind to receptors on the cell membrane, or enter the cell, and induce changes in cell metabolism that result in behaviors like cilia beating.

Approximately 550 million years ago (4 billion years after earth's formation), what is called "the Cambrian explosion" occurred. In a relatively short period, geologically speaking (13–25 million years,) virtually all major phyla of multicellular life forms appeared in the sea and began to evolve and differentiate.

What drove the Cambrian expansion is not known. One possibility is that some "extinction" event changed the climate to favor complex multicellular eukaryotic organisms, at least in some niches. However, some bacteria have never gone extinct. Bacteria constitute the most successful life forms on earth, both in terms of number of cells and total weight or volume. As long as

Right: *Sea life during the Cambrian era, c. 550 million years ago.*

WHY ARE THERE BRAINS?

Animals make rapid, complex, coordinated movements of specialized body parts regulated by external inputs acquired by specialized organs such as eyes. Plants, on the other hand, do not do these things. Although they may react to their environment on slower time scales, such as stomata opening and closing to allow gases to move in and out of their internal tissues in response to moisture and daylight, they lack the complex specialized organs of animals that make brains necessary. The evolutionary trajectory that caused animals to evolve brains to rapidly process sensory input and control movement eventually produced the most complex object we know of in the universe, the human brain.

Jellyfish use electrical signals to coordinate cells to allow them to propel themselves through the water.

Rhopalia

Nerve nets and circular muscles on subumbrella

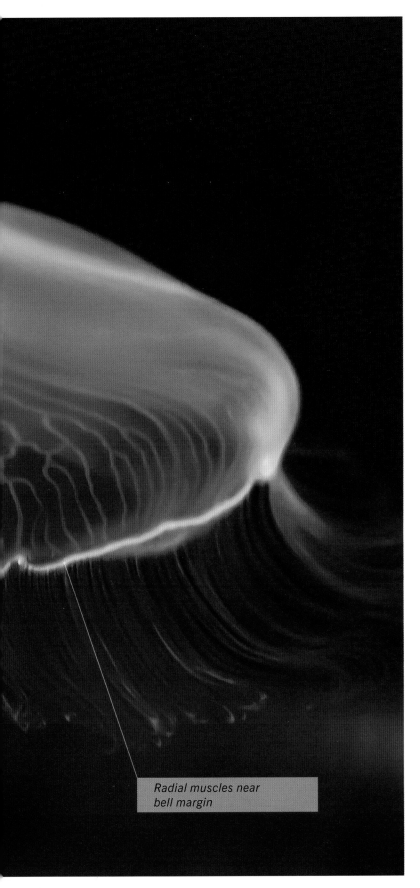

Radial muscles near
bell margin

there is any life on earth, there will be bacteria. Another hypothesis is that multicellular eukaryotic organisms that are dependent on high oxygen concentrations could not exist until the photosynthetic organisms had generated it.

The origins of primitive nervous systems

Multicellular organisms have specialized internal and external cell clusters that act as organs with particular functions. For example, cilia (microscopic hairs) on cells inside a multicellular group are not particularly effective in group motility, so cells on the outside become specialized for inducing motion. Other surface cells become specialized for sensing the environment, while internal cells take care of metabolic functions such as nutrient and waste processing. Specialized cells for sensing and movement are known as neurons. Some scientists believe that the first true eyes arose in the post-Cambrian period, causing the elaboration of complex ganglia (clusters of neurons) to process visual information to enhance survival. All the post-Cambrian phyla evolved ganglia that sensed the environment, produced movement, and coordinated sensation and action. Although plants later also evolved specialization between internal and external cells, only animals evolved nervous systems needed for fast coordination during movement.

Animals are classified into various groups according to their descent. All the phyla that arose at the Cambrian expansion are invertebrates, animals without backbones or spinal notochords, except one, the Chordata. The invertebrate phyla include worms, crabs, starfish, and later, insects.

The nervous systems of chordates were generally the most complex of all animal phyla, with some notable exceptions. The octopus has a major ganglion at the base of each of its eight arms to control the arm. There is also a central ganglion that communicates with all the arm ganglia, but the number of neurons in the central ganglion is far fewer than that in each of the arms. The total number of neurons in the nine octopus' ganglia is about 500 million. This is a very large number for an invertebrate, and octopuses are clearly very intelligent animals. For comparison, a fish brain has about 100,000 neurons, the frog's brain has about 16 million neurons, the rat brain, about 70 million. The human brain has about 86 billion neurons.

Effective movement requires coordination between cells that mediate motility. A major solution to this problem is electrical signaling between cells, which

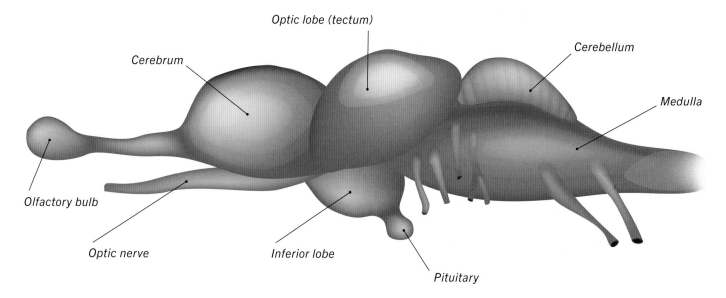

Labels on figure:
Optic lobe (tectum)
Cerebrum
Cerebellum
Medulla
Olfactory bulb
Optic nerve
Inferior lobe
Pituitary

Above: *Anatomy of a fish brain.*

makes them neurons. Electrical signaling forces a network of connected cells to behave in unison, such as the contraction of a jellyfish "bell" to propel it through the water. The jellyfish moves by activating contractile cells along the rim of its bell to push it through the water. These cells must contract at the same time to generate good thrust. Electrical coupling between the cells ensures that, as one cell initiates a contraction, the rest simultaneously also do so. This coordinated contraction is produced without a true central nervous system, but with small clusters containing hundreds of cells, called ganglia.

Chordates, vertebrates, and true brains

Most animals we are familiar with, including all vertebrates, are Chordata. Chordates are bilaterally symmetric and have a notochord (hence, chordate) of axons and ganglia running along their dorsal surface. Most early chordates were wormlike. Some of these wormlike precursors gave rise to more complex swimming animals that eventually led to all cartilaginous and vertebrate fish, then amphibians, reptiles, mammals, and, ultimately, humans. Large, anterior brains arose in all chordate lineages.

Brains allowed complex control of multiple appendages instructed by sophisticated sensory processing, such as the brain of a fish, a typical vertebrate brain. Most fish have strong olfactory and visual senses. The olfactory bulb receives inputs from external olfactory receptors and projects to what is called the cerebrum (the part of the brain which generates and manages all complex

information), believed by some brain scientists to be the precursor to the mammalian neocortex. A large optic lobe processes visual information, while the cerebellum coordinates fine movement. The medulla (see page 48) and other brainstem structures mediate communication between the brain and spinal cord and provide low-level control and processing. As we will see later (see pages 32–3), all vertebrate brains have similar structures upon which evolutionarily newer structures, such as the neocortex, are added.

In summary, neurons and brains arose to:

1. Sense the external world through modalities such as sight, sound, and smell.
2. Sense the internal milieu such as body temperature, carbon dioxide, and glucose concentration.
3. Control voluntary muscles.
4. Control fight or flight status and digestive systems.

The Cambrian explosion 550 million years ago started an "era" of earth's history called the Paleozoic that lasted for about 250 million years. Prior to the Paleozoic era only simple multicellular life existed, only in water, mostly in the oceans. The Paleozoic era saw the growth of plants that could survive on land as well as the appearance of animals on land, starting with invertebrates like worms and insects, then amphibians, and finally, early reptiles. The post-Cambrian era saw a significant rise in organism complexity, and, in chordates, the enlargement of the brain.

The Permian and later catastrophes

The Paleozoic era ended abruptly at the Permian extinction event about 250 million years ago, the most devastating event in the history of life on earth, with the demise of 95 percent of extant marine and 75 percent of extant terrestrial species. Hypotheses for the cause of this event are an extraterrestrial impact, or large-scale volcanism. After the Permian extinction came the Mesozoic era, known generally as the age of dinosaurs. This era is comprised of three "periods" called the Triassic, Jurassic, and Cretaceous. Extinctions less severe than the Permian ended the Triassic period, which was followed by the Jurassic period. The Jurassic period was ended by another extinction that brought about the Cretaceous. Little is known about these extinctions,

but it is possible that they were due to meteorite impacts or climate change caused by these impacts.

The Permian extinction produced another rise in animal complexity, particularly in the brains of land animals, the dinosaurs. Although dinosaur brains are smaller than current mammals of the same body size, they are larger than those of their precursor fish and amphibians. Most dinosaurs had brains like or larger than those of modern reptiles such as crocodiles, having in the order of 80 million neurons. Dinosaurs evolved in sophistication until the Cretaceous extinction (called the K-T boundary) that ended the dinosaur era and led to the Cenozoic period in which mammals arose. There is strong evidence that the Cretaceous extinction was caused by an asteroid impact off the Yucatan peninsula.

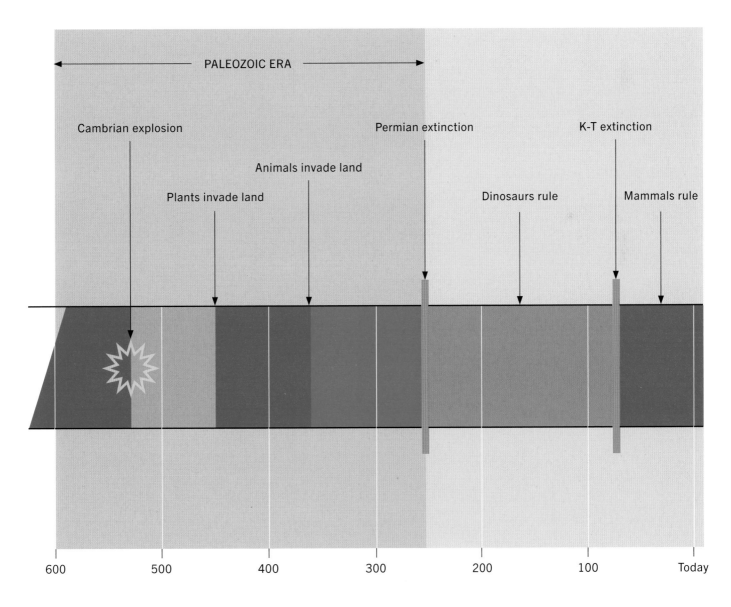

MILLIONS OF YEARS AGO

// Mammals and human brains

Some small, rodent-like animals that survived the Cretaceous extinction gave rise to mammals. Mammals enlarged their brains to diversify lifestyles to fill the terrestrial ecological niches left by the disappearance of the dinosaurs. Mammals differ from other, so-called cold-blooded vertebrates, such as reptiles and amphibians, in several notable characteristics, such as giving birth to live young, rather than laying eggs, and nurturing of young by lactation. These characteristics allow greater development of the brain in utero, and greater modification of the brain after birth through learning.

Development of the neocortex

The first mammals were rodent-like omnivores that depended highly on their sense of smell. The cerebrum in these species processed information from over 1,000 olfactory receptor types to make fine discriminations of odor information. Mammals then enlarged the cerebrum to become the neocortex that was then used to process visual and auditory information as they invaded niches where those senses were essential, niches that had become available with the extinction of the dinosaurs. Non-mammalian vertebrates have almost no neocortex, but in many mammals, the neocortex is so enlarged that it dwarfs the size of the rest of the brain and surrounds the rest of the phylogenetically older brain structures. In the simplest terms, the mammalian brain is a reptile brain embedded in a huge neocortex. The neocortex has taken over most sensory processing and motor control.

The neocortex changed the way that animals process sensory information and execute complex behaviors. Non-mammalian, cold-blooded vertebrate brains have brain nuclei that are phylogenetically very old and highly efficient and specialized for sensory processing and motor control in the niche the animal inhabits. But, with mammals radiating out into new niches, they needed to do three things not provided by the brains of their non-mammalian ancestors:

1. Expand brain tissue that could process sensory information with previously primitive senses with higher acuity than their predecessor brains would allow.
2. Expand brain tissue to allow the computational processing of new motor functions such as using the forelimbs for grasping.
3. Allow learning and plasticity in the brain to program it to survive successfully in these new niches.

The ability to learn via the neocortex also produced a greater capability for forming more complex social groups than previous non-mammalian vertebrate species. Social cues that were originally based on smell and simple visual signals were extended to more complex visual signals, such as facial expressions, and complex auditory signals, such as calls. Whereas most reptiles react to others of the same species along the two dimensions of male versus female, and bigger or smaller than the individual, some mammals

Below: *Brain evolution from the earliest structures of the fish brain to the complexity of the human brain.*

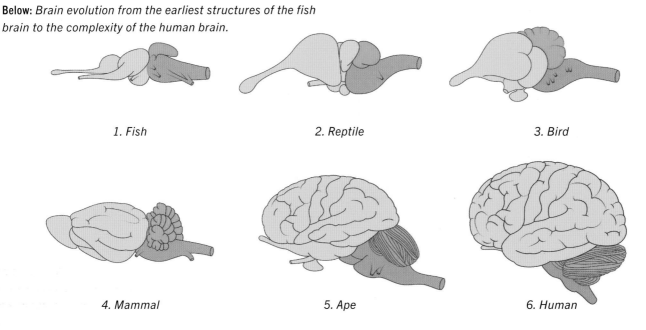

1. Fish   *2. Reptile*   *3. Bird*

4. Mammal   *5. Ape*   *6. Human*

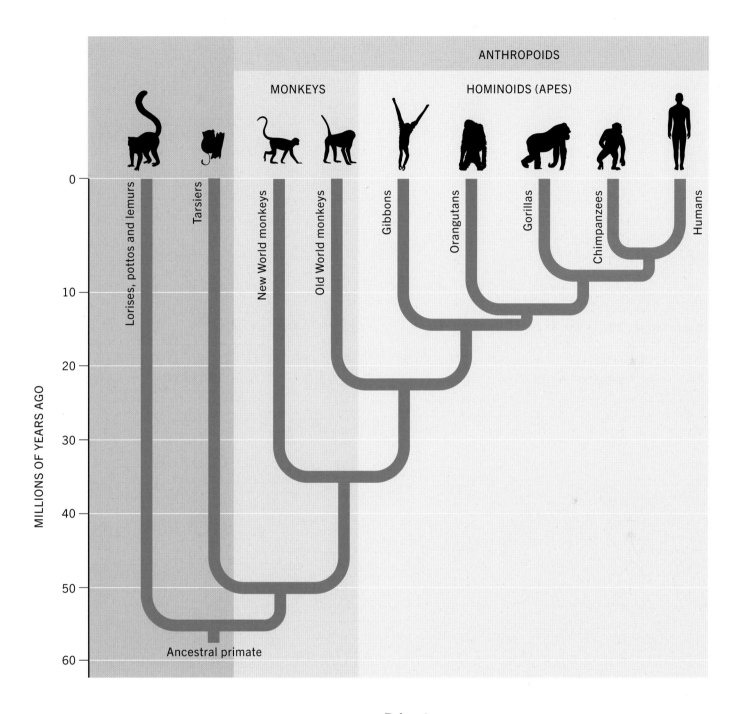

MILLIONS OF YEARS AGO

ANTHROPOIDS

MONKEYS

HOMINOIDS (APES)

Lorises, pottos and lemurs

Tarsiers

New World monkeys

Old World monkeys

Gibbons

Orangutans

Gorillas

Chimpanzees

Humans

Ancestral primate

form complex social hierarchies dependent on learning and behavior.

Thus, mammals used the general strategy of brain enlargement, mostly in the neocortex, to gain enhanced sensory processing, motor control, and learning, to adapt to new sensory niches. Today, mammals are either mostly visual (primates and predators), mostly auditory (bats and dolphins), mostly olfactory (dogs and cows), or mostly somatosensory via whiskers (voles, rats). Mammals exhibit extraordinary diversity in size and in the niches they inhabit, from arctic snowfields to the driest deserts on earth.

Primates

About 30–60 million years ago a mammalian order called primates began to diversify. These were mostly small tree dwelling mammals. Just as mammalian brains generally became larger than their reptilian predecessors to adapt to new niches, primate brains became the largest per body size of nearly all mammals, associated with adaptation to the new, highly 3D visual niche of tree dwelling.

The chart below shows the hypothesized evolutionary relationship and time of origin of currently existing primates. The closest non-primate relative of primates that may resemble the ancestral primate is thought to be the tree shrew, a squirrel-like tree dwelling insectivore now living in southeast Asia. From these arose tarsiers, lorises, pottos, and lemurs.

The enlargement of the neocortex in primates exceeded that in other mammals, particularly in the frontal lobe. Primates like chimpanzees and humans have larger brains than non-primate mammals like rats and cats, even as a percentage of body weight, but this enlargement is particularly notable in the frontal lobes, the most anterior part of the brain. As we will see later (see page 155), the frontal lobes are the seat of working memory and abstract planning that enable behaviors such as tool use and complex social interactions.

Homo sapiens and others

The homo line produced several species that arose at least 3.5 million years ago, as shown to the right. Species of the genus Homo are characterized by upright walking, tool use, and enlarged brains compared to other primates. Humans (*Homo sapiens*) are the only surviving members of the genus homo. Some of the known extinct homo species are shown to the right, along with their origin and extinction times. Note that, although we humans pride ourselves as being the most advanced of the homo genus, virtually all the other hominids except heidelbergensis lasted longer than we have so far. Homo erectus, for example, lasted for 2 million years, and remains of this now extinct species are found all over the world.

The genetic study of human origins argues for a revision of one of the classical ideas of evolutionary branching, which is that species bifurcate like branches of a tree and that once a species splits into two different species, the two branches never rejoin. But it is now known from genetics that virtually all non-African humans have 1-4 percent Neanderthal DNA. Also, a large percentage of current humans have 1–7 percent DNA from a recently discovered, extinct hominid called Denisovan.

Neanderthals and Denisovans died off after a mass extinction around 30,000 years ago. *Homo sapiens* rapidly evolved afterward to become the dominate primate species on earth. For this, they used their large brains capable of advanced sensory processing, motor control, and learning to inhabit new niches, and to adapt to climate-induced changes in the niches they already occupied. This dominance was accelerated and enabled by the invention of spoken language, hypothesized to have occurred about 200,000 years ago. Although many other mammals have a variety of calls, no other animal uses a complex syntax that allows for noun and verb phrases and the virtually infinite generative power of human language.

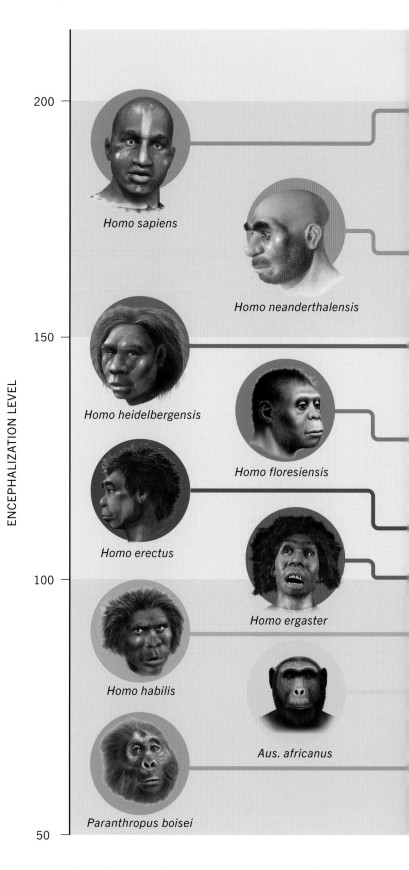

ENCEPHALIZATION LEVEL

200

Homo sapiens

Homo neanderthalensis

150

Homo heidelbergensis

Homo floresiensis

Homo erectus

Homo ergaster

100

Homo habilis

Aus. africanus

Paranthropus boisei

50

Above: *The encephalization level, or brain size, of the different hominid species and the time in which they lived.*

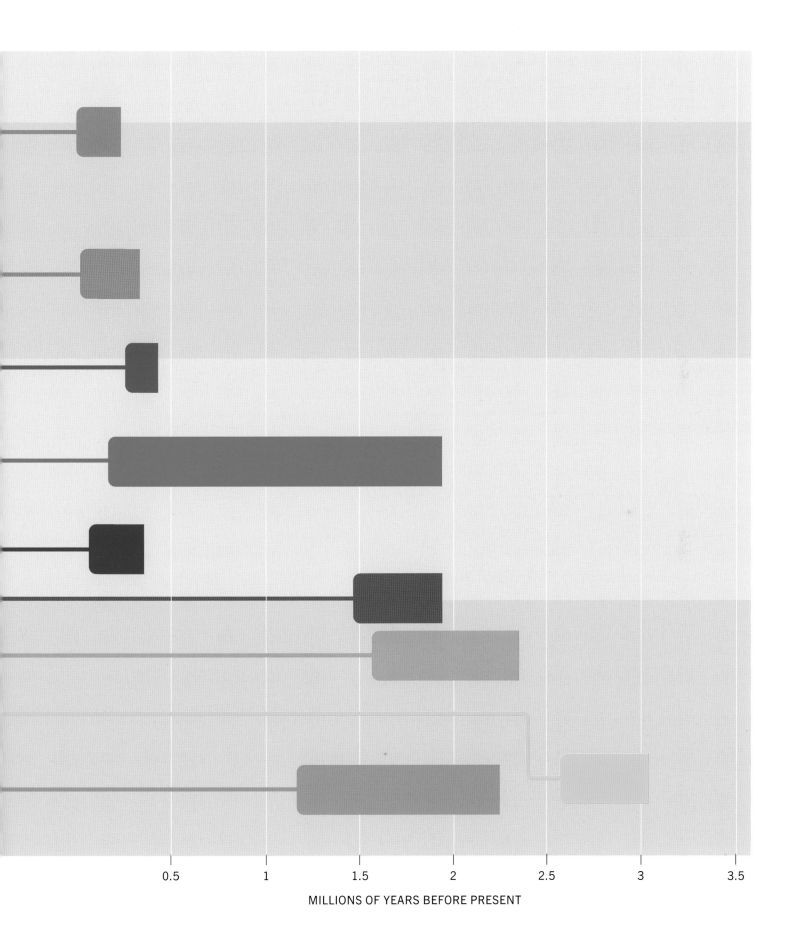

MILLIONS OF YEARS BEFORE PRESENT

// Development of the brain through the lifespan

The brain develops, like other body organs, from embryonic precursor cells. In the case of the brain, the precursor cells are from a layer called the "ectoderm." The major stages in brain development include the ectoderm cells forming what is called the neural plate, followed by the folding and fusing of the neural plate to form the neural tube. Next, the neural tube differentiates into a rudimentary brain and spinal cord. These structures then differentiate into areas of neural somas versus areas of axon tracts. Finally, synaptic genesis and pruning "wires" the neural connections of the central nervous system and its connections with the peripheral nervous system.

Fetal development

The fertilized egg develops into an organism of billions of cells in at least 78 organs each containing over 200 specialized types. The nervous system is the most complex array of cells in the body with the largest variability in cell types.

After fertilization, the first cell divisions of the ovum form a spherical structure called a blastula. The cells in the blastula then divide into three main cell lines in an inner versus outer onion-like pattern in a process called *gastrulation*:

1. The outermost *ectoderm* gives rise to the nervous system and skin.
2. The middle *mesoderm* produces muscle, bone, and connective tissue.
3. The innermost *endoderm* yields the digestive and respiratory systems.

Below: *The zygote divides into a "blastula" before dividing into three cell lines in the process of gastrulation.*

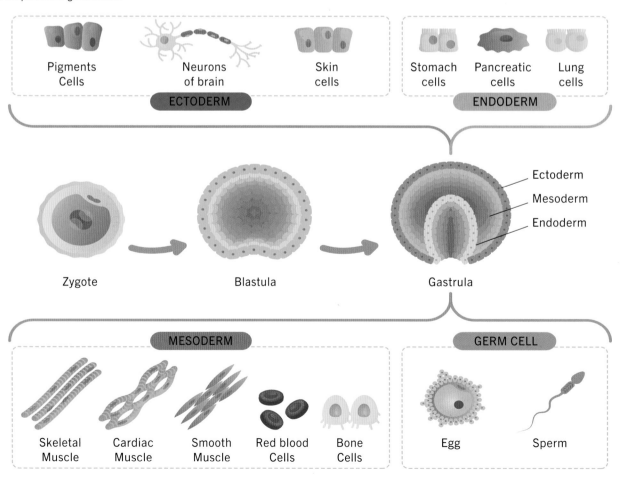

Pigments Cells   Neurons of brain   Skin cells   **ECTODERM**

Stomach cells   Pancreatic cells   Lung cells   **ENDODERM**

Zygote   Blastula   Gastrula

Ectoderm
Mesoderm
Endoderm

MESODERM   Skeletal Muscle   Cardiac Muscle   Smooth Muscle   Red blood Cells   Bone Cells

GERM CELL   Egg   Sperm

The three-layer gastrula then differentiates along two main axes via processes such as invagination and cell migration:

1. Anterior versus posterior.
2. Dorsal versus ventral.

The left and right halves of the body are roughly symmetrical, with some important exceptions. Cells in the ectoderm surround the entire developing embryo, whereas the mesoderm and endoderm layers become segregated dorsally versus ventrally inside the body.

The next landmark stage in nervous system development is neurulation, when the ectodermal cells on the dorsal surface form what is called the neural plate.

Once the anterior-posterior and dorsal-ventral axes are established, the central nervous system continues to differentiate into various portions of the brain and spinal cord.

After neurulation, embryogenesis proceeds by the main mechanisms of:

1. Neural proliferation.
2. Neural differentiation.
3. Neural migration.

These processes produce the major structures of the central nervous system, and the general internal organization of these structures, such as layers of various types of cell.

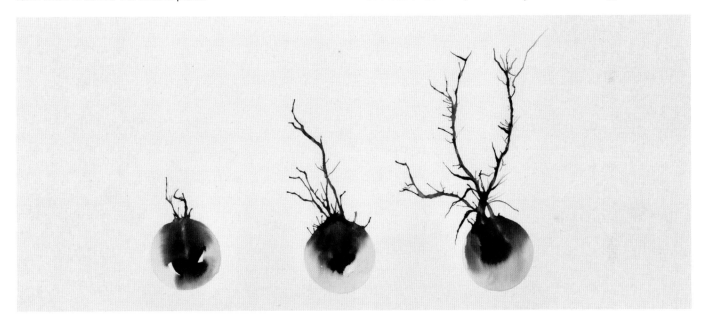

Above: *The development of neurons: axons grow out from the spherical cell body.*

Below: *Differentiation of the central nervous system over time.*

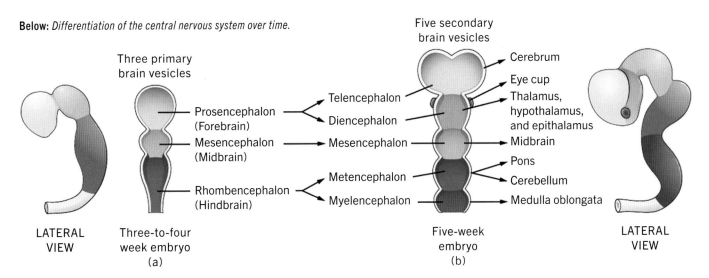

Birth and infancy

Three other processes that organize nervous system connections during fetal development also continue after birth, namely:

1. Synapse formation.
2. Synaptic pruning.
3. Myelination.

At birth, the number of neurons in the central nervous system is about the same as in early adulthood, but the complexity of connections and extent of myelination (the glial wrapping of axons that speeds spike transmission) is much reduced. The number of neurons changes much less than the complexity of the neurons' dendritic trees. This is a function of the number of cross connections between neurons, which increases dramatically in the first six months of life.

Synapse formation and pruning

Dendritic complexity is the result of the processes of synapse formation and pruning. A major difference between animals with large versus small brains is the extent of neural wiring determined by experience. In large-brained animals the nervous system initially elaborates many more synaptic connections during development than will survive into adulthood. Among a neuron's thousands of inputs, those that are activated at the same time as other inputs may contribute to the firing of the postsynaptic cell, while other inputs may not share this coincidence. Inputs whose activation is uncorrelated with the firing of the postsynaptic cell will be "pruned," according to the paraphrase of the Hebb principle, "cells that fire together wire together."

The Hebb mechanism allows the learning of associations. If every sheep you see growing up is white, neurons that fire for the sheep shape and those that fire for the white color will be preserved as inputs to some postsynaptic neurons that fire for white sheep, thus forming the neural substrate for this concept. The use of learning mechanisms such as coincident firing allows the nervous systems of large-brained animals to be vastly more sophisticated than the smaller nervous systems of small-brained animals programmed mostly by genetics. After all, in humans, there are only about 20,000 protein-coding genes, but our brains store hundreds of thousands (at least) of concepts. Moreover, the use of learning to program synaptic connections fine-tunes the organization of the brain to the specific environment encountered during development, which might range from that of an Inuit living on an Arctic ice sheet to that of a tribal member in the jungles of central Borneo.

Right: *Brain development in the early years of life.*

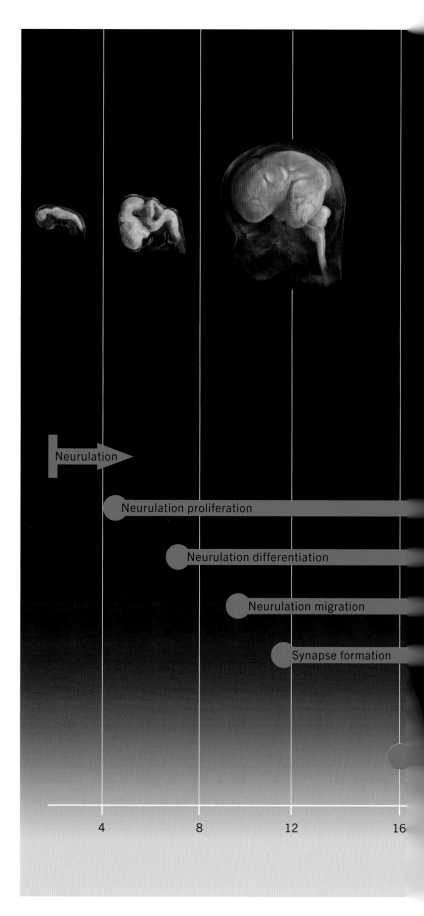

Neurulation

Neurulation proliferation

Neurulation differentiation

Neurulation migration

Synapse formation

4   8   12   16

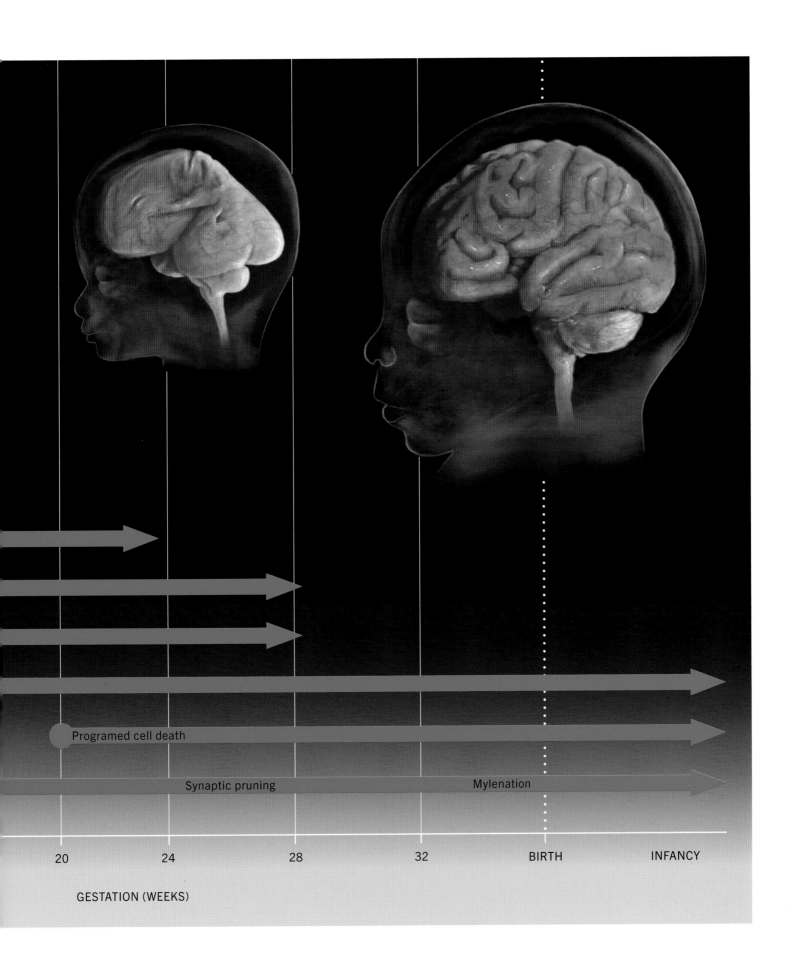

Programed cell death

Synaptic pruning

Mylenation

20   24   28   32   BIRTH   INFANCY

GESTATION (WEEKS)

During infancy, the number of neurons in the brain remains relatively constant. The major processes taking place include the fine-tuning of connections by mechanisms such as the Hebb process described above, and myelination. The fine-tuning of connections occurs in different regions of the brain at particular times, called *critical periods*.

Critical periods and synaptic competition

Critical periods are times during which synapse formation and pruning are particularly dependent on experience. After critical periods, which typically last months, synaptic plasticity is greatly reduced and there is reduced impact of experience on synapse formation. Critical periods demonstrate the importance of competition in synaptic pruning.

Myelination

Myelination is the other main process occurring during infancy and up to early adulthood. Myelination is the process by which certain types of glial cells wrap the axons of neurons everywhere except for certain gaps. The myelin sheath prevents the leakage of ionic currents between the gaps, so that action potentials, instead of being continuously conducted along the axon, jump from gap to gap in a process called salutatory conduction. This results in a 10-fold increase in spike speed along the axon.

Myelination is one of the longest duration processes in brain development. In humans, myelination of axon tracts in the frontal lobes is not completed until after teenage years. This lack of frontal lobe myelination during teenage years is correlated with reduced frontal lobe function in teenagers compared to middle age adults, such as tendencies toward impulsiveness and lack of planning.

Middle age

From the mid-twenties until a highly variable point in the sixties the brain is relatively stable in terms of neuron numbers, synaptic complexity, and myelination, outside of genetic abnormalities or injury due to environmental factors. Learning occurs mostly by strengthening active synapses and pruning ineffective ones. It has been shown recently, however, that new neurons are produced in adulthood in specific brain areas, such as the olfactory bulb and hippocampus.

Although reflexes and the speed of sensory processing and motor actions are declining in middle age, several factors appear to compensate for these declines, notably the increase in knowledge and experience during the lifespan,

Below: *From left to right, neuron development in a child, at 9 months, 2 years and 4 years.*

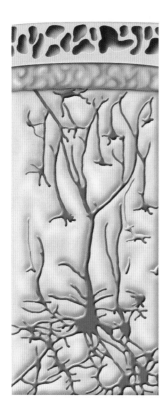
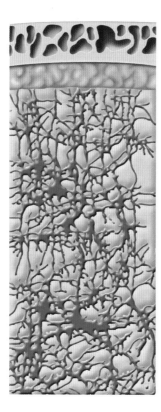
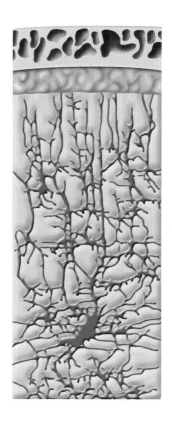

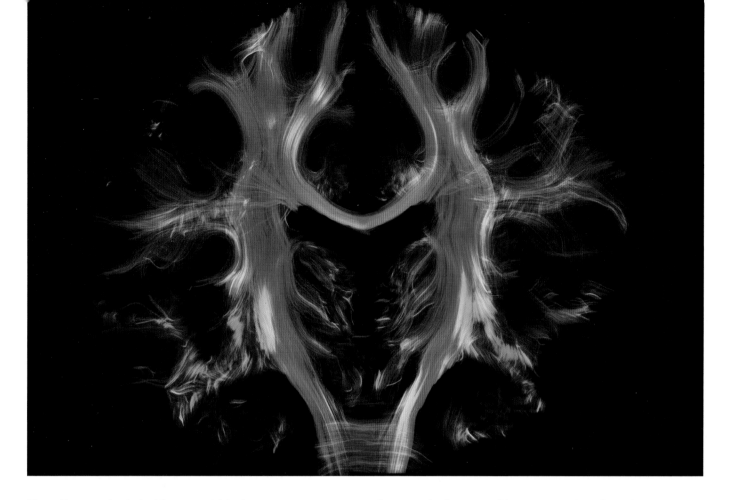

Above: *Tractography of a healthy young adult brain.*

typically subsumed under the idea of wisdom. The U.S constitution has minimum, but not maximum ages for elected representatives and the president. A recent dichotomization of young versus older problem solving strategies is that of fluid versus crystallized intelligence. Whereas younger brains are better at fluid intelligence, such as rapid processing of new information and linear reasoning, older brains are superior at case-based reasoning referencing a current problem to one about which one has stored knowledge.

Old age and cognitive decline

Eventually, however, enough synapses and neurons are lost during aging that accumulated knowledge cannot make up for the loss.

Actual decline in cognitive function depends on many genetic versus environmental variables, of course. Alzheimer's disease, for example, does not uniformly occur at a particular age, although its incidence increases monotonically as a function of age. Outside of pathologies such as Alzheimer's, education level and continued intellectual activity may prolong the ability to use crystallized intelligence to offset decline in neural function.

Exercise

Recent evidence has suggested that types of exercise that involve learning and cognitive decisions, like playing tennis,

for example, have cognitive preservative effects beyond those expected from simply maintaining fitness and vascular function, such as derived from jogging, for example. The learning component of skilled athletic activities appears to convey benefits similar to that derived from engaging in purely intellectual activities.

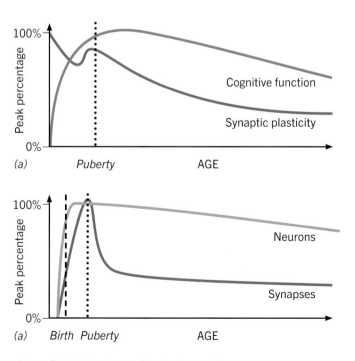

Above: *Graphs showing cognitive decline over time.*

// The nervous system

The central nervous system, includes the brain, spinal cord, and retina. The nervous system overall refers to the network of neurons that carry electrical messages to and from the brain and spinal cord to the rest of the body. The brain is just one component of the nervous system, though perhaps the most important. The nervous system is responsible for sensing the external world, regulating the internal milieu of the body, including factors such as body temperature, controlling the muscles and managing the digestive and adrenal systems. Without it, the complex interplay of animal life would not be possible.

Divisions of the nervous system

In order to understand how much of the body's neural activity of which we are not consciously aware influences consciousness, we look at the two major divisions of the nervous system:

1. The central nervous system (CNS), comprised of the brain, the retina of the eye, and the spinal cord;

2. The peripheral nervous system (PNS), containing neurons outside the central the central nervous system.

The peripheral nervous system has three main divisions that receive:

1. Sensory information from skin, joint, and muscle receptors, and motor control of muscles.
2. Sensory information from organs and glands in the "autonomic" nervous system and output control of those organs.
3. Sensory information from, and control of, the digestive tract in the "enteric" nervous system.

The autonomic nervous system controls the balance between organ homeostasis (the self-regulation of maintaining stable internal conditions such as temperature) needed for long-term survival, versus requirements for short-term action, needed for the fight or flight mode. The enteric nervous system is concerned mostly with the function of digestion.

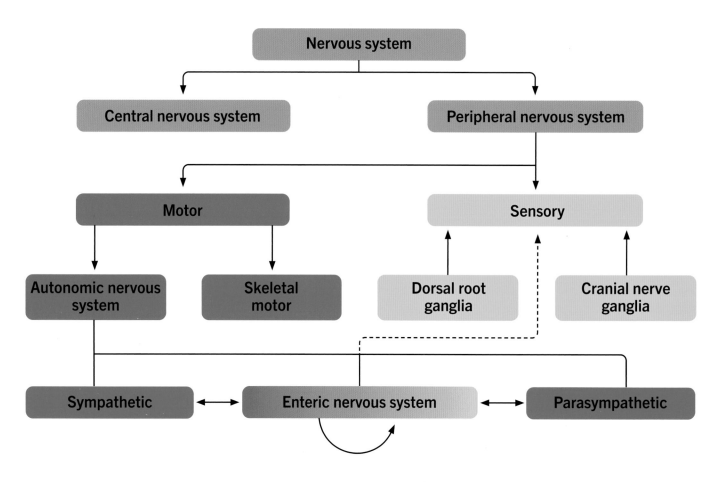

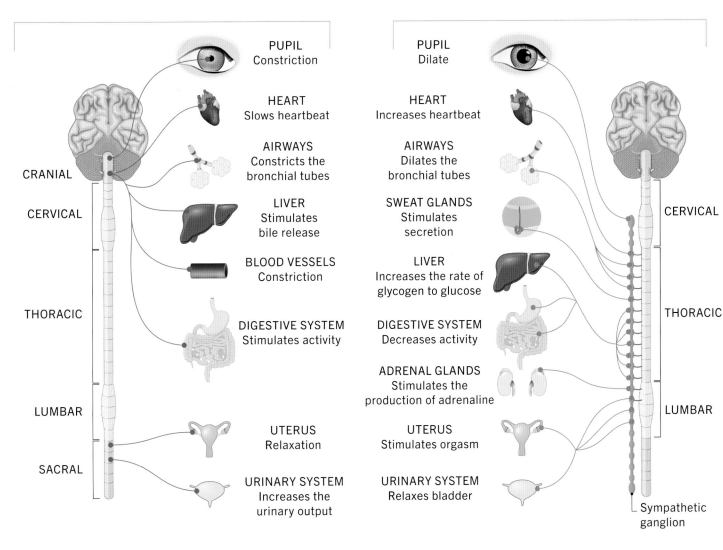

PARASYMPATHETIC

PUPIL
Constriction

HEART
Slows heartbeat

AIRWAYS
Constricts the
bronchial tubes

LIVER
Stimulates
bile release

BLOOD VESSELS
Constriction

DIGESTIVE SYSTEM
Stimulates activity

UTERUS
Relaxation

URINARY SYSTEM
Increases the
urinary output

CRANIAL

CERVICAL

THORACIC

LUMBAR

SACRAL

SYMPATHETIC

PUPIL
Dilate

HEART
Increases heartbeat

AIRWAYS
Dilates the
bronchial tubes

SWEAT GLANDS
Stimulates
secretion

LIVER
Increases the rate of
glycogen to glucose

DIGESTIVE SYSTEM
Decreases activity

ADRENAL GLANDS
Stimulates the
production of adrenaline

UTERUS
Stimulates orgasm

URINARY SYSTEM
Relaxes bladder

CERVICAL

THORACIC

LUMBAR

Sympathetic
ganglion

The enteric and autonomic nervous systems operate primarily unconsciously, but aspects of their operation can interact with the central nervous system and reach consciousness through effects on emotions.

The autonomic nervous system

The autonomic nervous system controls the operation of the organs of body sustenance and survival, such as the heart, lungs, and kidneys, along the axis of short-term versus long-term necessity. The sympathetic system deals with short-term needs — so-called "fight or flight" responses — by increasing heart rate and breathing and diverting blood flow to muscles and away from digestive organs. The parasympathetic nervous system deals with long-term needs and relaxing the body, returning it to a rest state — by

processing food, slowing heart rate and relaxing muscles. The autonomic nervous system receives sensory input from the central and peripheral nervous systems and has outputs to those same systems.

The *parasympathetic* system projects from the brainstem and sacral segments of the spinal cord to ganglia associated with various target organs such as the heart. These fibers use the neurotransmitter acetylcholine (ACh).

The *sympathetic* fibers originate in the thoracic and lumbar spinal cord and terminate in a set of ganglia just outside the spinal cord. Postganglionic neurons in those ganglia project to the target organ using the neurotransmitter norepinephrine. Thus, at the target organs, acetylcholine (parasympathetic) and norepinephrine (sympathetic) have opposite effects.

The autonomic nervous system regulates such systems as:

1. Breathing and heart rate.
2. Internal temperature.
3. Hunger and thirst.
4. Sleep-wake cycles.

Some organs that are under the direct neural control of the autonomic system also receive inputs from brain structures such as the hypothalamus via hormones released into the blood steam.

The sympathetic versus parasympathetic state of the autonomic nervous system is controlled by the brain and affects the brain. Although most autonomic functions take place unconsciously, autonomic function can rise to consciousness and even dominate it through the mechanism of emotion. The great American psychologist William James famously suggested that fear comes after the lower-level perceptual recognition of danger and even after the physical reaction to it. This is consistent with the neuroscientist Joseph LeDoux's idea of the "low road" pathway involving the amygdala (see page 142).

The vertebrate central nervous system

The central nervous systems of primitive, cold-blooded vertebrates such as fish and lizards (shown in red), consists primarily of:

1. A *spinal cord* that controls body movement and receives sensory information from the periphery.
2. A *brainstem* above the spinal cord, comprised of the midbrain, pons, medulla, and cerebellum (see pages 48–52). The brainstem integrates head senses such as vision, audition, and head orientation (vestibular), with spinal cord sensory and motor neurons to control behavior.

The earliest mammals added the *"limbic system"* (see page 53) (purple) above the brainstem. Limbic system structures such as the hippocampus, amygdala, thalamus, and cingulate gyrus allowed learning to modify instinctual and emotion-driven behaviors. Limbic nuclei receive information from most sensory areas of the brain that generates an "emotional state" that drives behavior that is dependent on past experience as well as current sensory information.

Later mammals such as primates and cetaceans, added the neocortex on top of the limbic system. The neocortex added higher acuity and generalization in sensation and allowed more complex motor programs for behavior.

Primitive vertebrate "lizard" brain

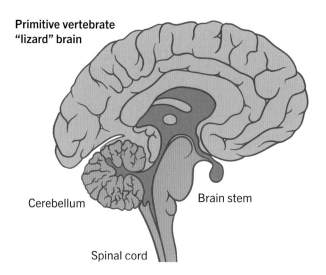

Cerebellum   Brain stem

Spinal cord

Primitive "mammal" brain

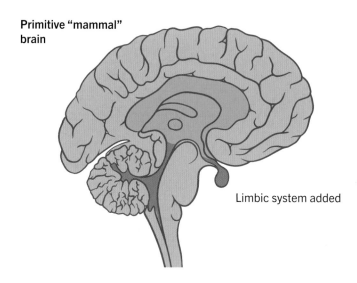

Limbic system added

Primate and high cephalic mammalian brain

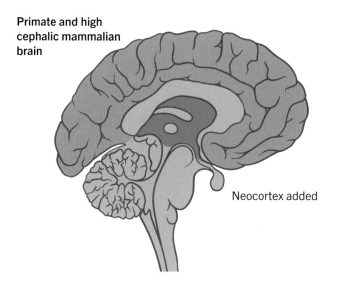

Neocortex added

Above: *Lizard, mammal, and primate brains.*

This evolutionary sequence created a central nervous system that is hierarchical in function. Many relatively automatic aspects of motor behavior can be completely controlled by spinal cord circuits. The limbic system added learning and some context-dependent control of behavior. The neocortex added sensory processing and motor control sophistication above the limbic system.

The spinal cord

The spinal cord is organized in segments. The number of spinal cord segments varies across vertebrate species. Humans have 31 segments of which there are eight cervical, twelve thoracic, five lumbar, five sacral, and one coccygeal. Similar sensory input from, and motor control of the face is via cranial nerves (called *corticobulbar*) from the brain.

Each segment of the spinal cord:

1. Receives sensory input from a particular area of the skin and joints enclosed by that skin area.
2. Projects motor output to control peripheral muscles at those joints.

Each spinal cord segment receives two major types of *sensory information* from the periphery:

1. Sensors in the muscles and tendons report limb position and muscle force.
2. Sensors in the skin report touch modalities such as pressure, vibration, temperature, and pain.

A cross section through a spinal cord segment reveals that it has a central gray area, where there are neural cell bodies and dendrites, surrounded by a peripheral white area of axons. Outside the spinal cord on the left and right sides are ganglia called dorsal root ganglia. The cell bodies of sensory neurons that receive input from the periphery are in the dorsal root ganglia. The cell bodies of lower motor neurons that project to muscles are in the front (ventral or anterior) spinal cord gray area.

Below: *The spinal cord.*

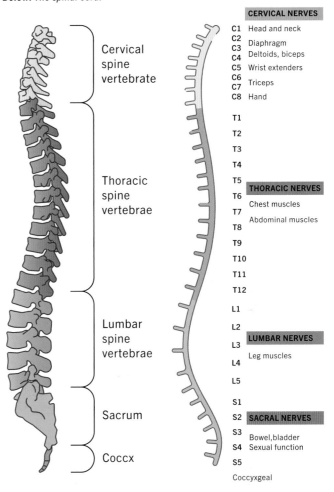

Cervical spine vertebrate

Thoracic spine vertebrae

Lumbar spine vertebrae

Sacrum

Coccx

CERVICAL NERVES	
C1	Head and neck
C2	Diaphragm
C3	
C4	Deltoids, biceps
C5	Wrist extenders
C6	
C7	Triceps
C8	Hand

T1
T2
T3
T4
T5

THORACIC NERVES	
T6	Chest muscles
T7	Abdominal muscles
T8	

T9
T10
T11
T12
L1
L2

LUMBAR NERVES	
L3	Leg muscles
L4	

L5
S1
S2

SACRAL NERVES	
S3	Bowel,bladder
S4	Sexual function

S5

Coccyxgeal

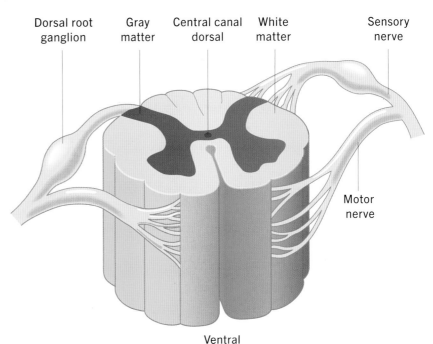

Dorsal root ganglion   Gray matter   Central canal dorsal   White matter   Sensory nerve

Motor nerve

Ventral

Right: *A cross-section of a spinal cord segment.*

Sensory input

Sensory information from receptors in the skeletal muscles, tendons, and joints is called *proprioception* (position sense) and *kinesthesia* (movement sense). This sensory information is used to modify the spinal cord motor output controlling muscle contraction to determine the motion of the limbs. Sensory information from pain, temperature, and some touch receptors is used to make limbs withdraw from dangerous contacts. The neurons that project sensory information have their cell bodies in the dorsal root ganglia that are located just outside the spinal cord on the left and right sides. The area of skin whose receptors send signals to one spinal cord segment is called a dermatome. Most skin sensory information is also relayed to the thalamus by sensory projection neurons. The thalamus then projects this sensory information to the somatosensory cortex in the brain.

Below: *Links of a spinal cord segment.*

Motor control

The spinal cord controls skeletal muscles via motor neurons. These neurons, called *lower motor neurons*, have their cell bodies in the ventral gray area of the spinal cord. Because the motor neuron axons go away from the spinal cord toward muscles on the periphery, they are called *efferents*. Lower motor neurons are considered part of the peripheral nervous system even though their cell bodies are in the spinal cord because they project to skeletal muscles outside the spinal cord.

Lower motor neurons receive inputs from and are activated by:

1. Proprioceptive input from the same half segment of the spinal cord.
2. Proprioceptive and interneuron input from the other half (left versus right) of the same segment of the spinal cord.
3. Neurons in other segments of the spinal cord.

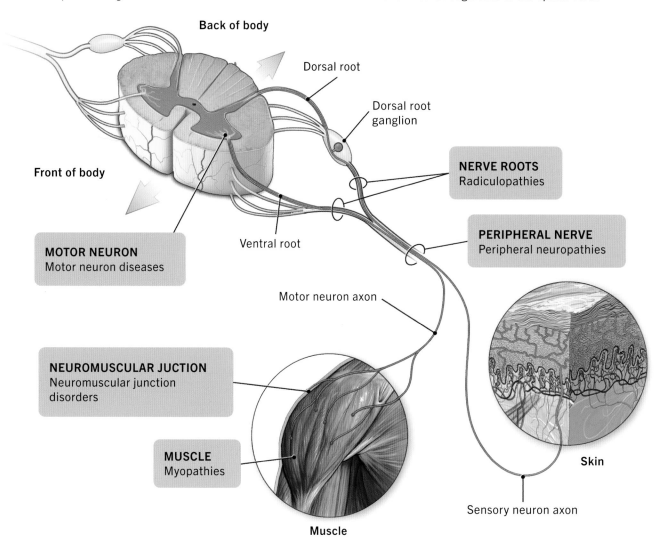

Back of body

Dorsal root

Dorsal root ganglion

NERVE ROOTS
Radiculopathies

Front of body

MOTOR NEURON
Motor neuron diseases

Ventral root

PERIPHERAL NERVE
Peripheral neuropathies

Motor neuron axon

NEUROMUSCULAR JUCTION
Neuromuscular junction disorders

MUSCLE
Myopathies

Skin

Sensory neuron axon

Muscle

Above: *Actions such as walking are controlled by motor neurons.*

4. Areas of the brainstem at the base of the brain just above the spinal cord (upper black arrows).
5. Areas of the neocortex called the motor cortex at the rear of the frontal lobe (upper black arrows)

The spinal motor output is coordinated by a range of motor control programs within the spinal cord, distributed across its segments along its length. Standing upright, and basic gaits like walking and running, can be executed completely by the spinal cord, without involving the brain. These spinal cord neural programs are called central pattern generators.

Complex spinal cord neural networks exist in primitive animals. The invertebrate octopus, for example, has more neurons in each of the eight ganglia that control its eight arms than it does in its central brain. Vertebrate brainstem nuclei add sensory information from cranial nerves, such as from head orientation (vestibular) and visual systems, for balance and obstacle avoidance. The primary motor cortex evolved to execute and learn new movement sequences.

The spinal reflex

A classic function mediated by spinal neural circuitry without the brain is the spinal reflex, typically tested as part of a standard physical examination. The spinal reflex shows the integrity of a complete behavioral circuit within the spinal cord, from stimulus to response. When the examiner taps the patellar tendon of the quadriceps muscle, it stretches the muscle, simulating what would happen if the knees were buckling (flexing) while standing up. The spinal cord has a reflex that compensates for this without brain involvement, in the following chain.

1. Stretch receptors in the muscle (muscle spindle) increase activity due to the stretch.
2. The afferent (moving towards) axons from these receptors carry this activity into the spinal cord.
3. In the spinal cord, these afferents activate lower motor neurons to the quadriceps muscle, causing it to contract and extend, reversing the stretch-flexure.
4. Simultaneously, the stretch receptor afferent axons activate inhibitory interneurons that reduce the force the opponent hamstring flexor muscle exerts so that its reflex does not counter the quadriceps extensor action.

This reflex happens before signals from the stretch receptor reach the brain. The brain can override this reflex with advance knowledge, of course. If you knew the doctor was going to do this test, you could send a prior brain signal before the patellar tap from your neocortex down the spinal cord to lock the flexor muscle to consciously block the action.

The motor control hierarchy in the central nervous system

When you walk, your brain does not get involved thinking about each specific step and muscle contraction in the walking sequence, but chooses direction and gait, and integrates information from the eyes and the ears. The lower motor neurons receive controlling inputs from brainstem nuclei in the midbrain and medulla, and from the primary motor cortex. The primary motor cortex also projects to the brainstem nuclei that project to motor neurons in the spinal cord.

The brain's contribution to walking is still necessary because the real world is different than walking on a treadmill. There are hills and holes. The brain determines the direction you are walking, how fast to walk, and what to do about obstacles ahead. Walking is not the right gait if there is a predator behind the next tree, requiring you to turn about face and run.

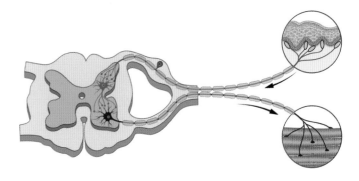

Above: *The spinal reflex arc.*

// The brainstem and limbic system

The portion of the brain, called the *brainstem* starts at the top of the spinal cord. This is nearly the entire brain of an early vertebrate such as a shark or lizard. The brainstem consists of three major subdivisions, starting from the top of the spinal cord: the medulla, pons/cerebellum, and midbrain.

The brainstem has three major functions: to control behavior using inputs from senses other than proprioception mediated by head organs, such as vision, audition, olfaction, taste, and vestibular; to fine-tune motor behaviors; and to integrate information from and control the autonomic and enteric nervous systems.

The medulla

The medulla is the lowest structure of the brainstem. It consists of several nuclei and fiber tracts. *Nuclei* are areas where there are clusters of cell bodies of neurons. The *nuclei* in the medulla make it markedly different from spinal cord segments. Some medullary nuclei receive inputs from higher brain centers or head sensory organs, process the inputs, and project them down the spinal cord or up to higher brain centers. For example, the *cochlear nucleus* receives acoustic input from the cochlea of the ears and projects this information upward to the brain.

The main *fiber tracts* running through the medulla are those whose axons are from neurons in the spinal cord, going to higher centers (nuclei) of the brain; or those from neurons in the brain going to the spinal cord.

Important tracts from the brain include:

- The *pyramidal* tract axons of upper motor neurons, beginning in the primary motor cortex and projecting through the brainstem to project lower motor neurons that drive muscles. In the medulla, these tracts cross from one side of the brain to the other, so that the left-brain hemisphere controls the right side of the body, and vice versa.
- The *tecto-spinal* tract receives visual information to coordinate head and eye movements by projecting to upper cervical segments that control neck muscles. When you turn your head, you are able to keep your eye fixed on a target thanks to this function.

Below: *The brainstem.*

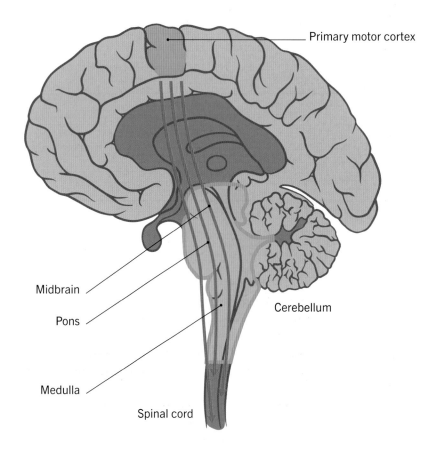

Primary motor cortex

Midbrain

Pons

Cerebellum

Medulla

Spinal cord

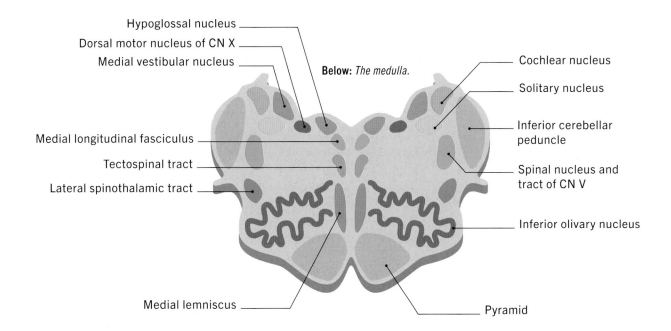

Hypoglossal nucleus
Dorsal motor nucleus of CN X
Medial vestibular nucleus

Below: *The medulla.*

Cochlear nucleus
Solitary nucleus
Inferior cerebellar peduncle

Medial longitudinal fasciculus
Tectospinal tract
Lateral spinothalamic tract

Spinal nucleus and tract of CN V

Inferior olivary nucleus

Medial lemniscus

Pyramid

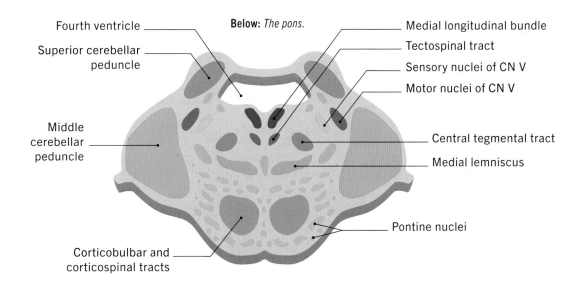

Fourth ventricle
Superior cerebellar peduncle

Below: *The pons.*

Medial longitudinal bundle
Tectospinal tract
Sensory nuclei of CN V
Motor nuclei of CN V

Middle cerebellar peduncle

Central tegmental tract
Medial lemniscus

Pontine nuclei

Corticobulbar and corticospinal tracts

The medulla thus receives inputs from and projects to the spinal cord and higher brain centers. It also has interconnections with the other two brainstem nuclei, the pons/cerebellum and midbrain.

The pons

The middle of the three brainstem structures is the pons and cerebellum. Axons project between the pons and cerebellum via large fiber tracts called cerebellar peduncles. Axons to and from the brain to the spinal cord traverse the pons white matter as in the medulla. Major tracts running through the pons include:

1. The middle and superior *cerebellar peduncles*. Peduncle is another name for tract. These tracts reciprocally connect the pons with the cerebellum.

2. The *corticobulbar* tracts receive inputs from face receptors much like the dorsal root ganglia at the spinal cord level.

3. The *corticospinal tract* (called "pyramids" at the medulla level) contain the axons of upper motor neurons that go from the primary motor cortex to synapse on lower motor neurons in the spinal cord.

Two notable nuclei within the pons are the sensory and motor nuclei of the trigeminal nerve (also known as *cranial nerve V*). This nerve carries touch, temperature, and pain sensations from the face to the pons. Motor neurons in this nucleus control facial muscles.

The pons also has several nuclei involved in the arousal and autonomic functions.

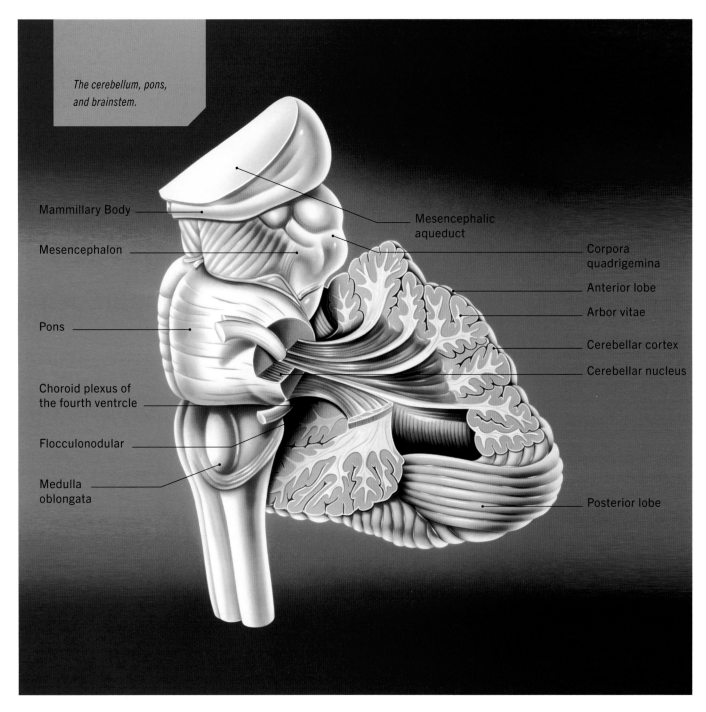

The cerebellum, pons, and brainstem.

Mammillary Body

Mesencephalon

Pons

Choroid plexus of the fourth ventrcle

Flocculonodular

Medulla oblongata

Mesencephalic aqueduct

Corpora quadrigemina

Anterior lobe

Arbor vitae

Cerebellar cortex

Cerebellar nucleus

Posterior lobe

The cerebellum

The cerebellum is a massive, complex neural center emanating from the rear (dorsal) side of the pons. Various estimates put the number of neurons in the cerebellum at over 50 billion, versus 86 billion for the entire brain. The three main functions of the cerebellum are:

1. Maintenance of balance and posture.
2. Coordination of voluntary movements.
3. Motor learning.

The cerebellum receives visual and vestibular information from higher brain centers and proprioceptive information from lower spine receptors via the pontine peduncular tracts. The cerebellum output projects to both upper motor neurons in motor cortex and lower motor neurons in spinal cord loops.

Opposite: *MRI image of the brain showing cerebrum, corpus callosum, and cerebellum.*

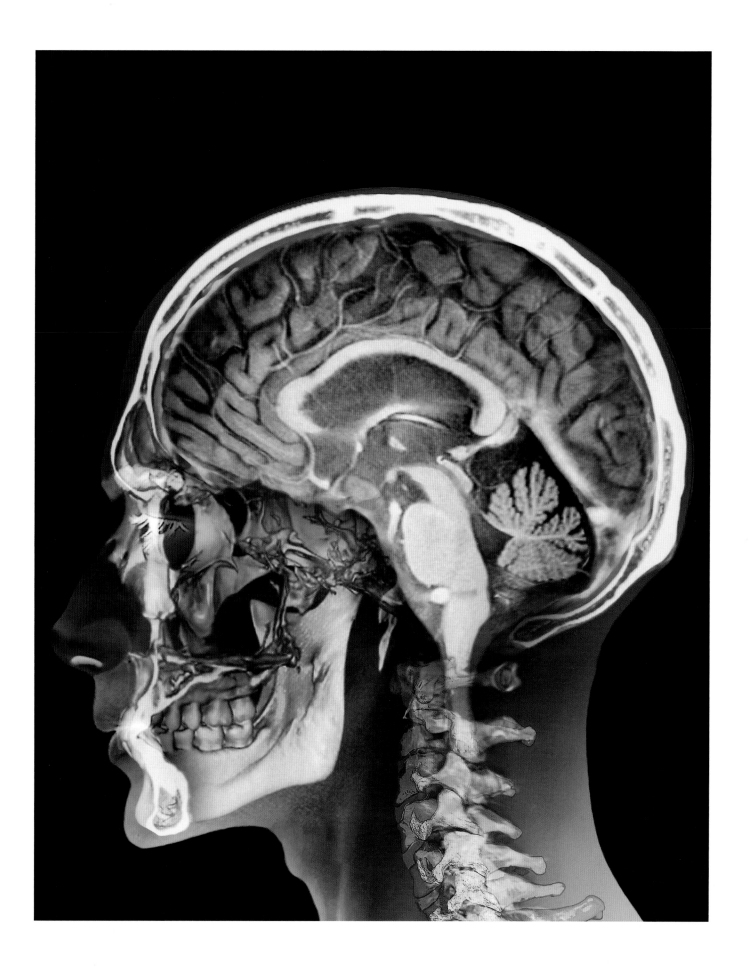

The midbrain

The same tracts that run to and from the spinal cord and brain run through the midbrain as well as the medulla and pons. Several important nuclei within the midbrain are important for vision, hearing, motor control, arousal, and temperature regulation.

In reptiles and amphibians, the midbrain is the main processing center for visual and auditory input. The midbrain transmits this information to the brainstem to control behavior.

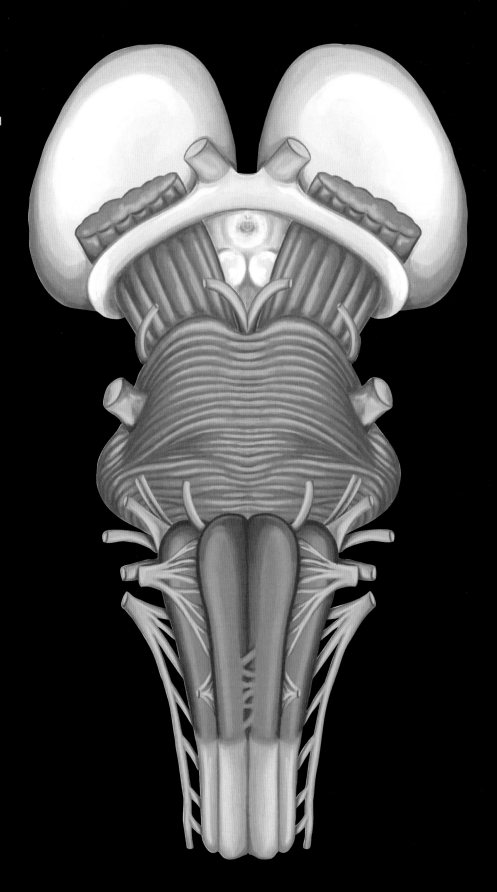

The limbic system

The first mammals added the limbic system on top of the brainstem. Three limbic structures of evolutionary significance in the transition from reptiles to mammals are the hippocampus, the amygdala, and the cingulate gyrus.

The hippocampus

The hippocampus is a crucial structure in learning, particularly, in learning based on complex sensory processing. In early mammals much of the input to the hippocampus probably came from the olfactory cerebrum and may have been involved in remembering routes for navigation.

In modern mammals, the hippocampus receives processed sensory input from virtually the entire sensory neocortex. Within the hippocampus, neurons with highly modifiable synapses, called NMDA receptors, form association circuits between arbitrary inputs. An example would be, that, at:

1. this time of day,
2. this olfactory signal,
3. this proximity to the water hole...

Some predator is likely to be around the water hole.

Learning via the hippocampus allows behaviors to be based on complex learning contingencies, rather than on pre-programmed sensory-triggered instincts only.

The amygdala

The amygdala performs a similar function as the hippocampus, but for a completely different purpose. The amygdala receives inputs from, and projects to, frontal areas of the brain that are important in social behavior. Animals like monkeys that live in complex social hierarchies need to be able to read facial expressions and remember the social status of many group mates, with disastrous aggressive or social outcast results if they don't. The amygdala mediates socially important learning that is necessary for operation in complex social groups—something that is much more common among mammals than reptiles.

The cingulate gyrus

The cingulate gyrus developed in early mammals, and ties together sensory processing, hippocampal and amygdala memory processes, and behavioral output. Its structure is what is known as mesocortex, which is older than true neocortex that evolved later. In larger mammals with a large neocortex, the cingulate gyrus acts as a command-and-control center that allocates attentional resources between various neocortical areas.

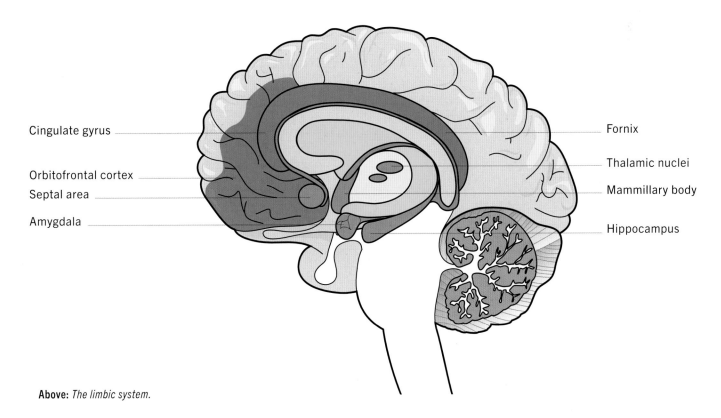

Cingulate gyrus

Orbitofrontal cortex
Septal area

Amygdala

Fornix

Thalamic nuclei

Mammillary body

Hippocampus

Above: *The limbic system.*

// Thalamus, neocortex, cortical lobes, and basal ganglia

The next structure up from the midbrain is the thalamus. The thalamus and neocortex together are brain structures added to the reptilian brain by mammals to enhance processing of sensory stimuli, and, with the basal ganglia, provide more complex control of goal-driven behavior. The thalamo-cortical system allows for a hierarchy of processing to exist for each sense. Receptors in the sensory periphery project (sometimes through intermediate nuclei) to the thalamus. That area of the thalamus projects to the newly evolved neocortex. In the neocortex are vertical neural circuits called columns. Each of these circuits can process a set of patterns from its inputs in a standard way. These columns then project to a higher level of neocortex that recognizes an even more complicated pattern from those inputs. This processing hierarchy can be extended indefinitely by enlarging neocortical area for that sense. Primates, for example, which are highly visual animals, have a hierarchy of over 30 neocortical areas specialized for visual processing.

Projections to the cortex pass through the thalamus

The thalamus sits at the center of sensor processing like the hub of a wheel. The senses of vision, audition, face sensation, and taste project to a specific area of the thalamus, which projects to a specific area of the neocortex. The sense of smell is a bit odd in that the olfactory lobe projects directly to a sensory processing area in the frontal lobe. However, even in olfaction, there is a pathway that includes the thalamus also.

Because the neural circuitry in the neocortex is similar everywhere within it, cortical processing happens similarly for all senses. A large percentage of neurons in the visual cortex respond only when an object in visual space is moving at a particular direction at a particular speed. In the auditory cortex, there are also found neurons that respond only to a sound moving at a particular speed, such as a chirping bird flying from one's left to right. Similarly, some neurons in the somatosensory cortex (which receives skin sensation) respond only to movement across the skin in a particular direction. When early mammals invaded ecological niches that required high acuity in visual, auditory, or somatosensory processing, they could use the same type of neocortex to enhance that sense.

Cortical lobes

The *occipital lobe* is almost entirely visual. The *parietal* lobe

Below: *The cortico-thalamic system.*

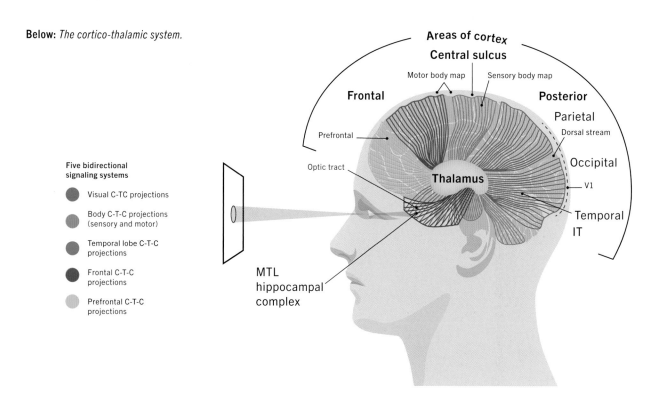

Five bidirectional signaling systems

- Visual C-TC projections
- Body C-T-C projections (sensory and motor)
- Temporal lobe C-T-C projections
- Frontal C-T-C projections
- Prefrontal C-T-C projections

Areas of cortex
Central sulcus
Motor body map
Sensory body map
Frontal
Posterior
Parietal
Prefrontal
Dorsal stream
Optic tract
Occipital
Thalamus
V1
Temporal
IT
MTL hippocampal complex

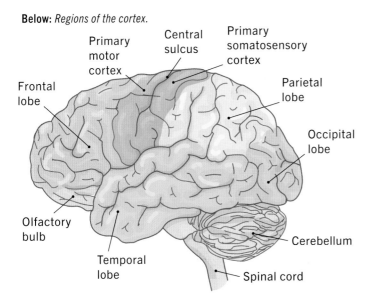

Below: *Regions of the cortex.*

Primary motor cortex

Central sulcus

Primary somatosensory cortex

Frontal lobe

Parietal lobe

Occipital lobe

Olfactory bulb

Cerebellum

Temporal lobe

Spinal cord

does somatosensory processing, as well as high-level visual and audio processing. Audio and high-level visual processing is done in the temporal lobe. The *frontal* lobe relates to motor functions, with the most abstract representation of motor goals located in the prefrontal cortex. The final upper motor neuron output is from the primary motor cortex, at the border of the parietal lobe somatosensory area. There are small areas of the frontal lobe that process some olfactory and taste inputs.

The basal ganglia and frontal lobes

Just as the thalamus and cortex evolved in mammals to form complex representations in various senses, so the basal ganglia arose in conjunction with the frontal neocortex for complex motor skills. Mammals with larger brains such as primates have an enlarged prefrontal cortex. In humans, this area is large even compared to other primates. The prefrontal cortex contains the highest-level representation of a goal. For any goal, there are many possible means of achieving it, and for each step along the way, many possible ways of achieving any of those means.

The action path from general goal in the prefrontal cortex to specific

motor output in the primary motor cortex is selected by the basal ganglia. The basal ganglia are a set of nuclei and tracts below the neocortex, just lateral to the thalamus (like the thalamus, there is one on each side of the brain). The basal ganglia receive sensory information about the positions of the limbs in relation to the body, and the position of the body in space, relative to some movement goal. The basal ganglia process this information and project to a motor area of the thalamus that projects to and controls the sequence of motor commands from the prefrontal cortex to the motor output in primary motor cortex. The sensory input may signal that the current method must be modified, or even that a switch must be made to a different method to accomplish the sequence.

The basal ganglia-cortex system in the frontal lobe proceeds from general goals activated in prefrontal areas, to select among different ways of achieving those goals, to the exact muscle contractions needed in each step toward those goals. Suppose the goal is to buy an extension cord to operate one's computer. This may itself be a sub-goal of using the computer to study for an exam. The frontal lobe stores various ways known to get the extension cord, such as driving to a store, taking a bus to the store, and so forth. The basal ganglia, "knowing your car is in the shop and the bus is running on time," might decide you should take the bus to the store. There are then many ways to take the bus, such as walking to the bus stop, taking Uber to the bus stop, and so forth. Other data allow the basal ganglia to choose one of those means. The end of all this is the primary motor cortex that commands particular muscles to make you move in a particular way.

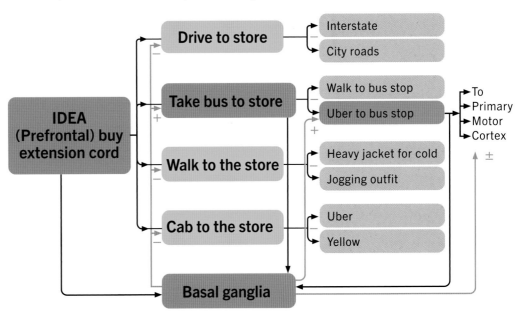

Above: *The decision-making tree.*

// How neurons work

As multicellular animals became larger, cells in the center became isolated from the periphery and information from receptors on cells in the periphery. Cells in different parts of the periphery had trouble executing coordinated action. The solution to these problems is the neuron. Without neurons, cells can only process information by relatively slow chemical reactions within themselves, and can only communicate with other cells via slow, hormone-like release of bio-chemicals. Neurons, in contrast, use much faster electrical communication within themselves, and between neurons. This section is about how they do this.

Neurons are cells with specialized structures

Neurons are cells whose cell bodies (called *somas*) have a nucleus with DNA, surrounded by cytoplasm containing metabolic machinery such as mitochondria (which produce energy). They have two major structures not found in most other cells, namely, dendrites that receive inputs to the neuron, and an *axon* that transmits the output of the neuron.

Neurons communicate by synapses

A *synapse* is a connection between the axon of one neuron and the dendrite of another (usually). The transmitting, or presynaptic neurons sends voltage pulses called *spikes*, or *action potentials* along its axon. At axon terminals, and sometimes other places on the neuron, there are either electrical or chemical synapses. Electrical synapses occur between adjacent neurons via *gap junctions* (see page 61).

Below: *Three different types of neurons.*

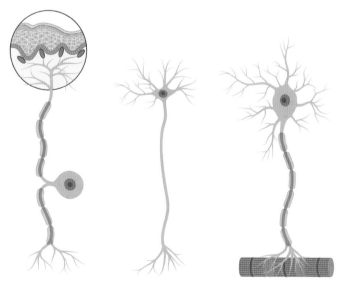

Sensory neuron   Interneuron   Motor neuron

In chemical synapses, when the action potential reaches the end of the axon, called the *axon terminal*, molecules of *neurotransmitter* are released into the gap between the two neurons, called the *synaptic cleft*. These neurotransmitter molecules diffuse from the presynaptic cell, releasing across the synaptic cleft to the membrane of the receiving, or postsynaptic neuron.

The sequence of neurotransmission consists of four major steps:

1. The axon terminal of the presynaptic neuron releases neurotransmitter into the synaptic cleft.
2. Neurotransmitter molecules bind to neurotransmitter receptors on the postsynaptic neuron.
3. Time-varying binding of all the receptors throughout the neural dendritic tree of the postsynaptic neuron causes time-varying electrical currents to flow to the soma of the neuron and the *initial segment* of its axon.
4. The initial segment of the axon produces a stream of voltage pulses called spikes or *action potentials* that travel along the neuron's axon to the next neuron.

Neurons typically have thousands of time-varying synaptic inputs on their dendrites and soma. These inputs can be excitatory, tending to make the receiving neuron more electrically positive, or *inhibitory*, tending to make it more negative. Excitatory neurons pass their signal on to other excitatory neurons, while inhibitory neurons tell other neurons not to fire.

The time-varying summation of these currents at the initial segment of the axon where it leaves the soma produces a time-varying train of action potentials that then move down the axon to other neurons.

Neurotransmitter release

The major steps in *neurotransmitter release* are:

1. An action potential (voltage pulse) reaches the axon terminal after traveling the length of the axon from the presynaptic neuron's soma.
2. The voltage pulse opens calcium channels in the axon terminal membrane.
3. Calcium flowing into the presynaptic terminal causes packets of neurotransmitter molecules (called *synaptic vesicles*) to fuse with the external membrane of the presynaptic neuron and dump the neurotransmitter molecules into the synaptic cleft (the gap between the pre- and postsynaptic neurons).

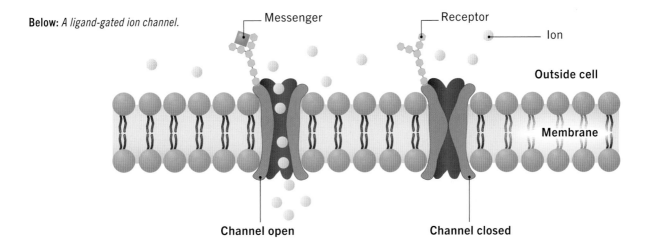

Below: *A ligand-gated ion channel.*

Messenger

Receptor

Ion

Outside cell

Membrane

Channel open

Channel closed

4. The neurotransmitter molecules diffuse across the synaptic cleft and bind to postsynaptic neuron receptors for those molecules.

The response of the postsynaptic cell to the release of neurotransmitter molecules depends on these molecules binding to receptors in the postsynaptic membrane that open *ion channels*. Ion channels are protein complexes inserted in the cell membrane that selectively allow some ions to move through the membrane when the channel is open.

Ion channels are embedded in the cell membrane. This membrane is a *lipid bilayer* consisting of a double array of molecules called *phospholipids*. These membranes do not permit the flow of water or most ions through them. Neurons insert selective ion channels into their phospholipid membranes that allow selective ions to traverse the membrane. Different types of ion channels can be either always open (*non-gated*) or gated. Gated channels can be opened by:

1. Voltage across the membrane (*voltage-gated*).
2. Binding by neurotransmitter molecules, called generically ligands (*ligand-gated*)
3. Other mechanisms such as mechanical stretch and pH.

Most ion channels are ion selective. Major ion channel types are selective for either sodium, potassium, chloride, or calcium. Some ion channels also allow the selected ion type to move in only one direction, either into the cell, or out of the cell only. These are called *rectifying channels*. Ion channels are the major means by which neurons generate fast electrical signaling.

The concentration of ions inside the cell differs from that outside. This is because neurons have in their membranes structures called *sodium-potassium pumps* (also called *transporters*) that create concentration differences in sodium and potassium inside versus outside the neuron.

Below: *A sodium-potassium pump.*

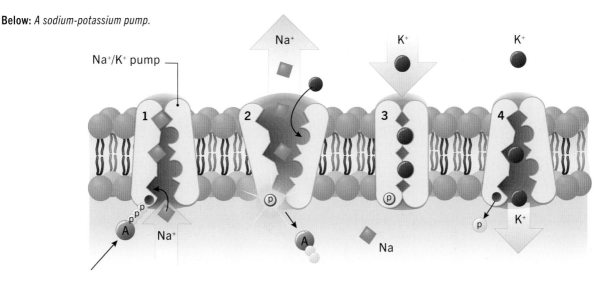

Na$^+$/K$^+$ pump

Na$^+$

K$^+$

K$^+$

1

2

3

4

Na$^+$

Na

K$^+$

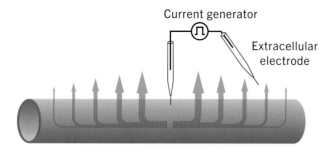

Current generator

Extracellular electrode

Above: *Current decays across membrane as distance increases.*

Sodium-potassium pumps run continuously in all neurons, removing sodium (Na+) from the interior of the cell, and bringing potassium (K+) in from the extracellular fluid, using the energy molecule ATP to move these ions against their concentration gradients. These pumps act in concert with ion leakage (non-gated) channels to produce two results:

1. Produce concentration differences in sodium and potassium to allow current flow through the membrane.
2. Cause the interior of the neuron to have a negative charge, or potential, with respect to the outside. This is called the *resting potential*, and for most neurons it is about -65 mV (millivolt).

The problem (and solution) of distance

So far, we have a picture of a neuron whose dendritic tree is receiving thousands of time-varying neurotransmitter inputs, producing a flow of currents throughout it. Some of this current reaches the soma and initial segment of the axon. But this leaves us with a problem. Many neurons transmit messages to other neurons a long distance away. For example, upper motor neurons in the brain send their axons through the brainstem and down the spinal cord to synapse on lower motor neurons several feet from the brain. While the passage of electrical current from dendritic synapses to the neural soma over distances of a few millimeters is efficient, transmission over distances of several feet is more problematic.

Below: *The action potential.*

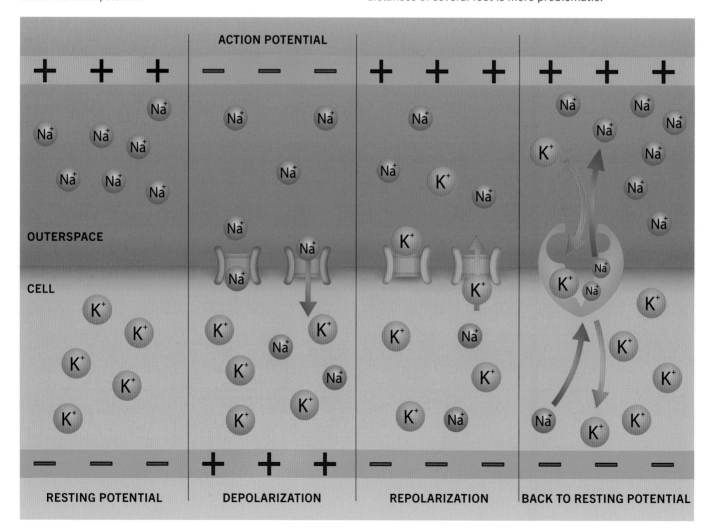

ACTION POTENTIAL

OUTERSPACE

CELL

RESTING POTENTIAL   DEPOLARIZATION   REPOLARIZATION   BACK TO RESTING POTENTIAL

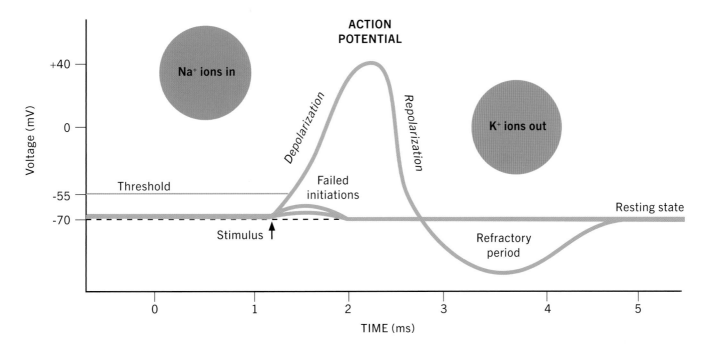

ACTION POTENTIAL

Voltage (mV)

+40

0

-55 — Threshold

-70

Na⁺ ions in

Depolarization

Repolarization

K⁺ ions out

Failed initiations

Resting state

Stimulus ↑

Refractory period

TIME (ms)

0   1   2   3   4   5

Above: *During the refractory period, it is difficult to generate another spike.*

Current flowing across the membrane is progressively lost the farther away one measures it from the injection site. This loss is due to:

- Axial resistance to current flow.
- Leakage of current across the membrane.
- The capacitance of the membrane smoothing or blurring the sharpness of the pulse at the injection site.

The solution is the *spike*, or *action potential*. Axons have ion channels for sodium (and potassium) that are voltage-gated. If the sum of all currents generated by the presynaptic inputs depolarizes the initial segment of the axon sufficiently, the voltage-gated channels for sodium will open. This leads to a chain reaction, as each subsequent segment of the axon is depolarized, all the way to the axon terminal several feet away. The depolarization level that produces the all-or-nothing spike event is called the *threshold*, which is around -55 mV in most neurons. The duration of a neural spike is about one millisecond.

Voltage-gated sodium and potassium channel currents bring about the action potential. The voltage change across the membrane in the neuron during an action potential is in blue. The rapid opening and closing of the sodium channels brings positive current into the cell, causing the inside voltage to be positive instead of negative. The slower potassium channel negative current (positive potassium going out) helps bring the membrane potential back to the resting potential.

After the voltage spike of the action potential, it is harder to generate another one for several milliseconds. This is known as the refractory period.

SUMMARY OF NEURAL OPERATION

1. Action potential pulses arrive at the axon terminal of the presynaptic neuron.
2. This spike rate is translated into a calcium concentration inside the axon terminal.
3. The calcium concentration is translated into a released neurotransmitter concentration in the synaptic cleft between the pre- and postsynaptic neurons.
4. The neurotransmitter concentration is translated into some percentage of open channels in the postsynaptic membrane.
5. The open channels at all the synapses on the postsynaptic neuron is translated into a current density in the initial segment of the axon.
6. The current density in the axon initial segment is translated into an action potential firing rate.
7. The action potentials travel unattenuated to the axon terminals of this postsynaptic neuron.
8. The action potential rate at the axon terminals of this neuron starts the chain at step 1 in the next neuron.

// Synapses

Neural communication is the process by which electrical changes in one neuron cause electrical changes in another. This occurs via synapses between the neurons. There are two major types of synapses: electrical and chemical.

Chemical synapses have much more flexibility than electrical ones and are by far the most common synapses in vertebrates. Electrical synapses have important functions in the central nervous system, however.

Synapses are responsible for communication between neurons, using either chemical or electrical signals.

Electrical synapses

Electrical synapses occur between adjacent neurons via gap junctions. *Gap junctions* are found not only between neurons, but also between cells in the heart, liver, and even in plants. At electrical synapses, current flows from an active neuron to an adjacent neuron through gap junction channels that allow ions and small molecules to move between the two cells. Gap junctions can be so closely packed that they constitute a significant percentage of the membrane contact area between the two cells.

Gap junction synapses are made by the mating of *hemi-channels* in the pre-and postsynaptic cells. The hemi-channel in each cell is formed by a six-subunit cluster of protein complexes called *connexins* that forms what is called a *connexon hemi-channel*. Each connexin subunit generally has four membrane-spanning domains.

Electrical synapses are fast because the flow of electrical current is nearly instantaneous, whereas chemical synapses take on the order of a millisecond for the presynaptic cell to activate the postsynaptic one. Electrical synapses are frequently found in circuits responsible for escape behaviors in invertebrates, where the activity of multiple postsynaptic target cells is synchronized. In vertebrates, electrical synapses are often found between inhibitory neurons in the brain. Even so, electrical synapses make up a minority of the synapses in the brains of mammals.

Electrical synapses have the inherent problem that the current injected through the gap junctions from one cell is diffused across all the postsynaptic membrane of the postsynaptic cells. This means that a single presynaptic cell can only drive one or a few smaller postsynaptic cells effectively, because there is no amplification. Thus, electrical synapses typically only provide synchronization, or low-dimensional modulation of activity.

Below: *Signal transmission at an electrical synapse.*

Chemical synapses

Although electrical synapses are fast, chemical synapses have several advantages of their own:

- They provide both excitation and inhibition.
- They allow amplification.
- They allow activity-dependent changes in synaptic efficacy.

Chemical synapses operate by having the presynaptic cell release a chemical messenger that binds to a receptor protein complex on the postsynaptic cell. This binding either directly or indirectly causes ion channels in the postsynaptic cell to open.

Chemical synapses fall into three main categories of chemical messengers:

1. **Neurotransmitters** act at synapses on postsynaptic neurons contacted by the axon terminal of the presynaptic neuron and produce rapid electrical changes in the postsynaptic neuron.
2. **Neuromodulators** produce slower, but often more long-lasting effects on postsynaptic neurons, in some cases distant from the presynaptic release site.
3. **Neurohormones** travel long distances in the body, often via the vasculature, to their final targets.

Signaling at chemical synapses occurs in two major steps. First, the presynaptic cell releases the neurotransmitter chemicals. Then, the specific receptors in the membrane of the postsynaptic cell react to the released neurotransmitter binding.

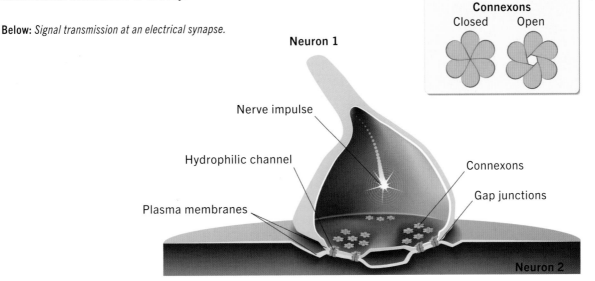

THE MAJOR STEPS OF NEUROTRANSMISSION

The major steps of neurotransmission

1. The arrival of an action potential at the presynaptic terminal causes voltage-dependent calcium (Ca 2+) channels there to open.
2. The calcium that enters the presynaptic neuron causes the fusion of synaptic vesicles containing neurotransmitter with the presynaptic terminal.
3. The neurotransmitter molecules released into the synaptic cleft (the gap between pre- and postsynaptic neurons) diffuse across and bind to receptors on the postsynaptic cell. The gap, or synaptic cleft between the presynaptic and postsynaptic neurons at a chemical synapse is about 20 nm wide, whereas the electrical synapse gap is only about 3.5 nm.
4. The receptor responses may be excitatory or inhibitory depending on whether the channel opened fluxes sodium (excitatory), or chloride or potassium (inhibitory).

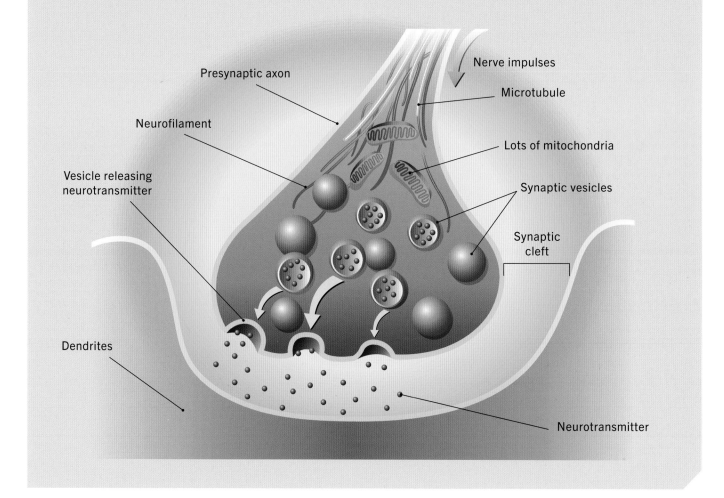

Presynaptic axon

Neurofilament

Vesicle releasing neurotransmitter

Dendrites

Nerve impulses

Microtubule

Lots of mitochondria

Synaptic vesicles

Synaptic cleft

Neurotransmitter

Molecules of neurotransmitter are enclosed in synaptic vesicles prior to release. These vesicles are anchored by proteins on the presynaptic membrane near release sites.

Calcium releases the vesicles from the protein anchors and stimulates the fusion between the membrane of the vesicle and the membrane of the axon terminal, dumping the neurotransmitter into the synaptic cleft. This process is called exocytosis.

The process of vesicle fusion continuously adds membrane to the presynaptic terminal. To offset this, areas of the terminal just outside the synaptic cleft undergo a process known as pinocytosis, where a vesicle-sized portion of membrane is "pinched off" and internalized into the cell.

Although the primary function of synaptic release is to affect another neuron, many synaptic terminals have *autoreceptors* for the neurotransmitter they release, just outside the synaptic cleft. These receptors provide feedback to the releasing neuron about its overall level of activity. Homeostatic mechanisms within the presynaptic neuron, with other factors, then modify the rate of synthesis and the release of neurotransmitter.

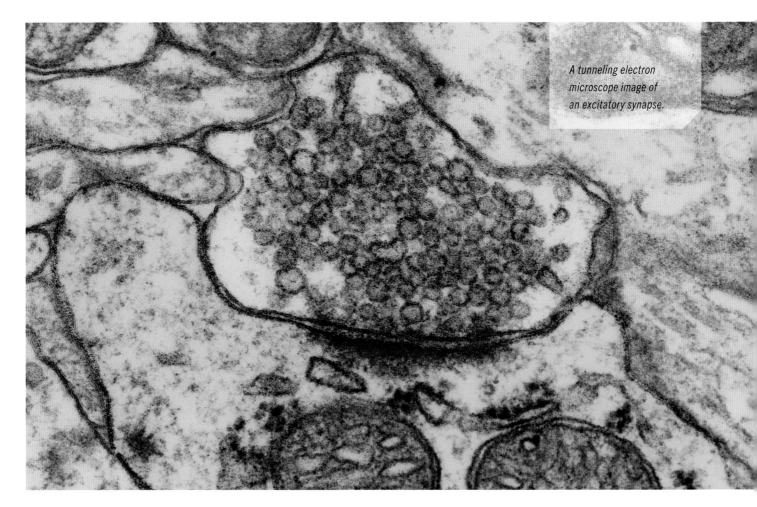

A tunneling electron microscope image of an excitatory synapse.

Postsynaptic receptors

The neurotransmitter molecules released into the synaptic cleft diffuse over to receptors in the cell membrane. These receptors have recognition sites on the protein receptor molecule complexes that bind only specific neurotransmitter molecules in a lock-and-key manner. Molecules that bind postsynaptic receptors are generically called ligands.

When the neurotransmitter molecule binds the ionotropic receptor protein, the protein undergoes a structural transformation that opens the ion channel. Most neurotransmitter receptors are ion-selective for either sodium, potassium, or chloride. Sodium channels are excitatory, whereas those for chloride and potassium are inhibitory.

The concentration of neurotransmitter molecules in the synaptic cleft declines by three different mechanisms:

1. Diffusion of the released neurotransmitter molecules out of the synaptic cleft.
2. Active enzymatic degradation of the neurotransmitter molecules.

3. Reuptake of the neurotransmitter molecules by the presynaptic terminal or by astrocytes.

The receptor protein complexes that open an ion channel in response to binding neurotransmitter molecules that we have discussed so far are called ionotropic. A second receptor type is called metabotropic. Metabotropic receptors do not have an ion channel within the protein complex. Rather, the binding of the neurotransmitter on the extracellular side of the complex causes them to release what is called a g-protein inside the postsynaptic cell. The freed g-protein initiates a biochemical cascade inside the neuron that opens nearby ion channels from the inside.

Metabotropic receptors are usually more slowly acting than ionotropic ones, with modulatory, rather than rapidly acting effects. Metabotropic receptors have wide-ranging effects on neurons through activation of a variety of what are called second messengers, whose activation may include the opening of ion channels, but also long-lasting effects due to changes in gene expression that can modulate synaptic efficacy or even delete or create new synapses.

// Neurotransmitters

Neurotransmitters that act at synapses fall into two major classes:

1. Small-molecule transmitters.
2. Neuropeptides.

Small-molecule transmitters rapidly activate postsynaptic responses, whereas neuropeptides are modulatory. Small-molecule transmitters are synthesized or repackaged in the axon terminal, whereas neuropeptides are synthesized in the cell body and packaged into vesicles that are transported the length of the axon to the axon terminal. Vesicles containing neuropeptides are used only once, whereas the vesicles are recycled with small-molecule neurotransmitters. Some axon presynaptic terminals can release both a rapidly acting neurotransmitter, and a more slowly acting neuropeptide.

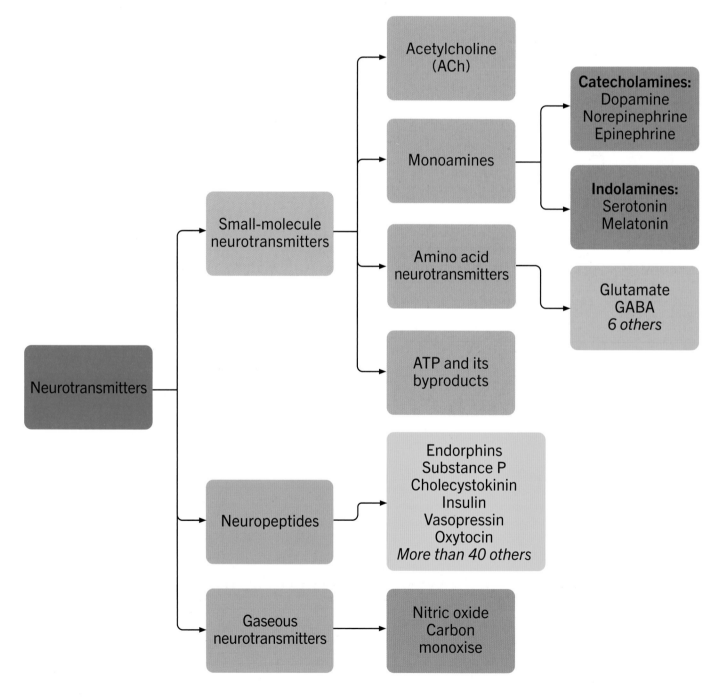

	Small-Molecule Transmitters	Neuropeptides
Synthesis	In axon terminal	In cell body
Recycling of Vesicles	Yes	No
Activation	Moderate action potential frequency	High action potential frequency
Deactivation	Diffusion, enzyme degradation, re-uptake	Diffusion only
Function	Fast neurotransmission	Neuromodulation

Small-molecule transmitters

These include:
1. acetylcholine,
2. five monoamines,
3. amino acids
4. ATP and occasionally its degradation byproducts.

Acetylcholine

Acetylcholine is a unique small molecule neurotransmitter that is not an amino acid or monoamine. Neurons that release acetylcholine (ACh) are referred to as cholinergic. ACh is the neurotransmitter of vertebrate motor neurons at the neuro-muscular junction. In the autonomic nervous system all preganglionic synapses use ACh as their neurotransmitter. ACh is also the transmitter at postganglionic synapses in the parasympathetic division of the autonomic nervous system.

Cholinergic neurons continuously synthesize the enzyme acetylcholinesterase (AChE) and release it into the synaptic cleft, where it breaks down ACh liberated from the presynaptic terminal. The choline resulting from this enzymatic breakdown is recaptured by the presynaptic neuron (reuptake) and resynthesized into more ACh.

Below: *The brain pathways of four important neurochemicals.*

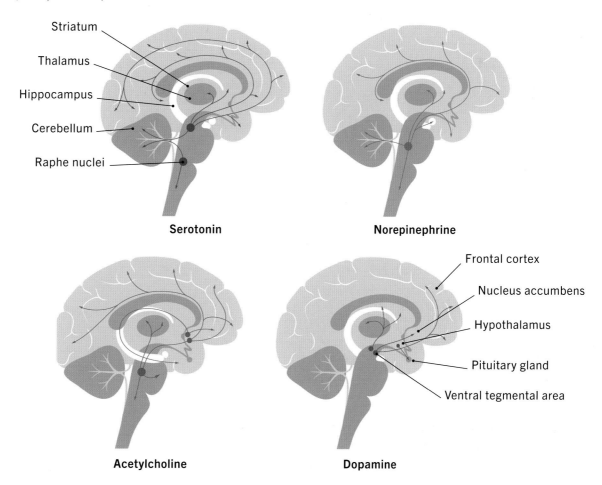

Striatum
Thalamus
Hippocampus
Cerebellum
Raphe nuclei

Serotonin

Norepinephrine

Acetylcholine

Dopamine

Frontal cortex
Nucleus accumbens
Hypothalamus
Pituitary gland
Ventral tegmental area

There are both ionotropic and metabotropic subtypes of cholinergic receptors for acetylcholine, known as nicotinic and muscarinic, respectively. These names come from exogenous substances. Thus, a nicotinic receptor responds to ACh released normally by the presynaptic terminal, but it also will respond to nicotine, as found in tobacco, applied exogenously. Muscarinic ACh receptors similarly normally respond to endogenous ACh, but they will also bind the exogenous substance muscarine, derived from the hallucinogenic (and poisonous) mushroom fly agaric.

Nicotinic receptors are fast ionotropic receptors, whereas muscarinic receptors are slower metabotropic receptors. Therefore, for example, you will find nicotinic receptors at the neuromuscular junction, for rapid muscular contractions.

Cholinergic neurons are widely distributed in brain regions such as the basal forebrain, septum, and brainstem that project to the neocortex, hippocampus, and amygdala. These cholinergic neurons participate in learning and memory. Cholinergic neurons are among the first to widely deteriorate in Alzheimer's disease, disrupting these functions.

Amino acid neurotransmitters

The most important small-molecule transmitters in the brain are the amino acid neurotransmitters. These can be divided into transmitters that are usually excitatory or inhibitory.

Excitatory amino acid neurotransmitters include:
1. L-Glutamate,
2. L-Aspartate,
3. L-Cysteine, and
4. L-Homocysteine.

Inhibitory amino acid neurotransmitters include:
1. GABA (gamma-aminobutyric acid)
2. Glycine,
3. β-Alanine, and
4. Taurine.

Glutamate

Glutamate is the most frequently used excitatory neurotransmitter in the central nervous system. Once presynaptically released, glutamate is taken up by both neurons and astrocytes to clear the synaptic area of excess glutamate, because extended action of glutamate on neurons can be toxic due to overstimulation.

There are three major types of ionotropic glutamate receptors named after the external substances that also activate them:

Right: *NMDA, AMPA, and GABA receptors.*

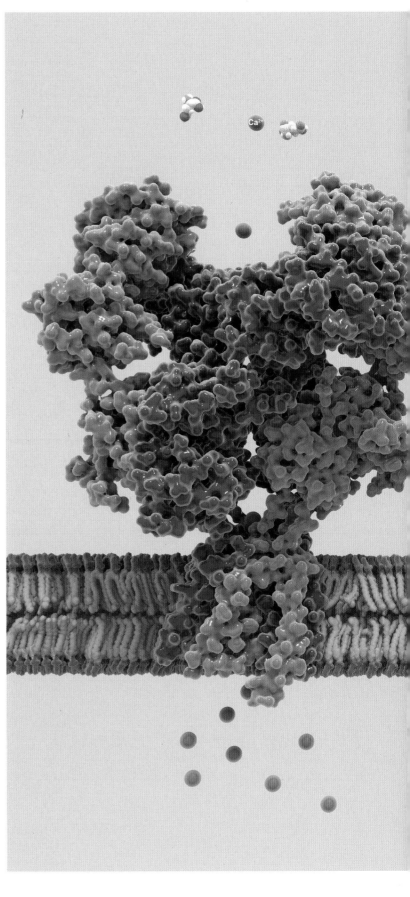

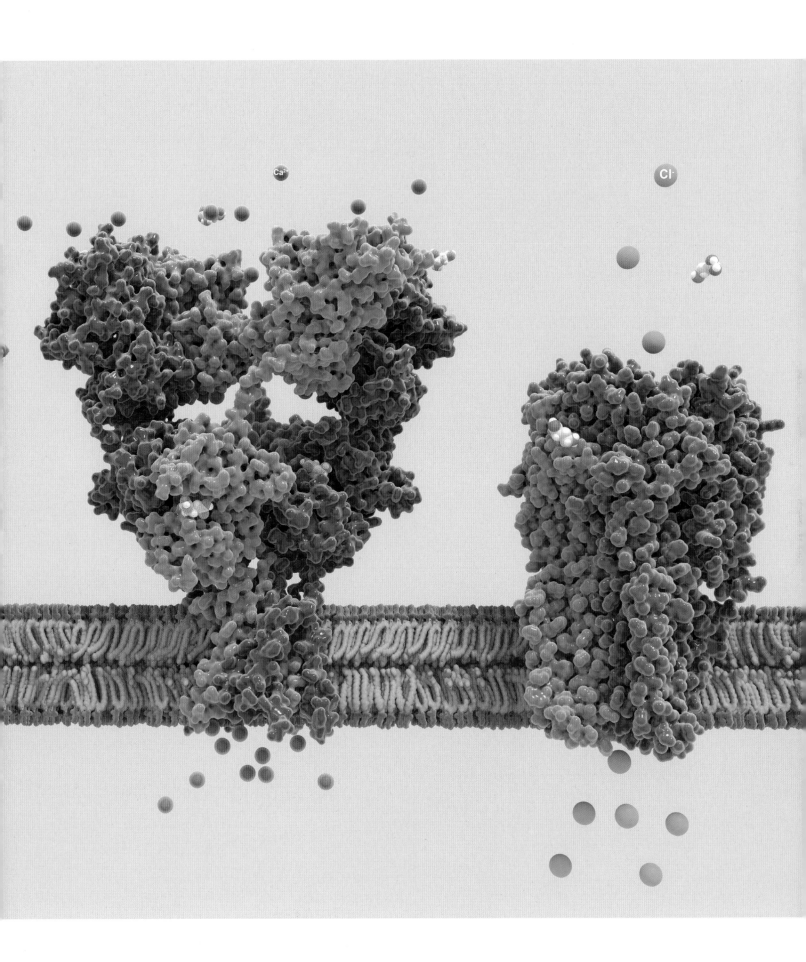

Above: *Glutamate is the most prevalent neurotransmitter in the brain and plays a key role in learning and memory.*

1. The N-methyl-D-aspartate (NMDA) receptor,
2. The alpha-amino-3-hydroxy-5-methylisoxazole-4-proprionic acid (AMPA) receptor, and
3. The kainate receptor.

The AMPA and kainate receptors open sodium channels when they bind a molecule of glutamate and are the most common excitatory receptors in the brain. The NMDA receptor also opens a sodium channel, but this channel also fluxes calcium, which has important modulatory effects on the postsynaptic neuron.

NMDA receptors are both voltage-dependent and ligand-dependent. They will not open unless glutamate is present and the postsynaptic membrane is depolarized at the same time. NMDA receptors are involved in coincidence detection and associative learning because they operate as AND gates: that is, two different inputs must be present to open the ion channel: (1) direct receipt of glutamate from the presynaptic part of the synapse, and (2) depolarization of the postsynaptic cell from a nearby excitatory synapse. The density of NMDA receptors is high in the hippocampus where associative memory formation occurs by detection of coincident inputs. NMDA receptors also allow the flow of calcium ions. Calcium ions activate enzyme sequences in the postsynaptic cell that can result in lasting structural and biochemical changes in neural functions, such as increasing synaptic efficacy. However, excess calcium entering the cell can be toxic. Neural damage such as during a stroke can cause a feedback loop in which the damage causes neural depolarization that releases excess glutamate that over-stimulates NMDA receptors, causing excess calcium entry.

GABA

GABA is the major inhibitory neurotransmitter of the central nervous system.

There are two major types of GABA receptors:

1. $GABA_A$ receptors are ionotropic chloride channels.
2. $GABA_B$ receptors are metabotropic receptors that open potassium channels.

The main effect of opening ligand-gated chloride channels is to clamp the neuron near the resting potential, opposing any depolarization induced by excitatory synapses.

GABA$_A$ receptors have additional binding sites besides the main channel opening site, called cofactors. These modulate the kinetics of the GABA$_A$ channel opening. Several of these sites are of clinical significance, such as sites that bind alcohol or barbiturates.

GABA$_B$ receptors are metabotropic receptors that, through second messenger intracellular cascades, open potassium channels. Opening ligand-gated potassium channels produces net inhibition.

Glycine

Glycine is another inhibitory neurotransmitter in the central nervous system, especially in the retina, spinal cord, and brainstem. Typical glycine receptors are ionotropic chloride channels. The toxin strychnine is a strong *agonist*

at ionotropic glycine receptors. Glycine is also a required co-agonist along with glutamate for NMDA receptors.

Monoamines

The five monoamine neurotransmitters are in two subgroups:

1. the *catecholamines* (dopamine, norepinephrine, and epinephrine) and
2. the *indolamines* (serotonin and melatonin).

All presynaptic neurons that release monoamines reuptake the neurotransmitter from the synaptic gap following release.

Dopamine in the brain

Dopamine is widely distributed throughout the brain and has modulatory functions for movement, reinforcement, and

Below: *Dopamine plays an important role in reinforcing reward pathways in the brain and is involved in addiction.*

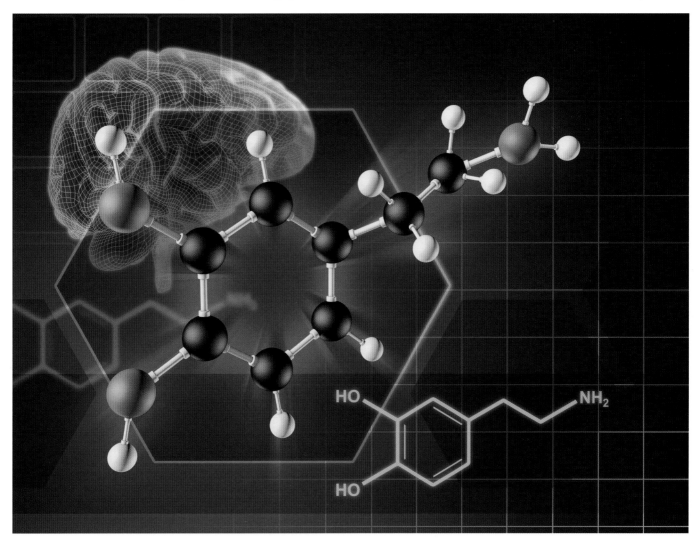

planning. Three major dopamine pathways originate in the midbrain:

1. Projections from the substantia nigra of the midbrain to the basal ganglia of the cerebral hemispheres provide an important modulation of motor activity. This pathway is damaged in Parkinson's disease (see pages 128–9), in which patients have great difficulties initiating and reprogramming movement.
2. The mesolimbic system arises in the ventral tegmentum of the midbrain and projects to various parts of the limbic system, including the hippocampus, the amygdala, and the nucleus accumbens. This system mediates feelings of reward and plays an important role in addiction.
3. Dopaminergic neurons in the ventral tegmentum project to parts of the frontal lobe of the cerebral cortex where they are involved in cognitive functions, particularly the planning of behavior.

The receptor subtypes for dopamine, D1 through D5, are all metabotropic. D2, D3, and D4 receptors can be found as both conventional postsynaptic receptors and presynaptic autoreceptors. Autoreceptors allow the presynaptic neuron to monitor its release of neurotransmitter. The D2 receptor class is involved in reward feelings and is an important target of psychoactive drugs.

Epinephrine and norepinephrine

Epinephrine and norepinephrine are commonly known as adrenalin and noradrenalin, respectively. Neurons releasing *epinephrine* are called adrenergic and those releasing norepinephrine are noradrenergic. Stress results in the release of epinephrine from the adrenal glands located above the kidneys into the blood supply. Neurons that secrete *norepinephrine* are found in the locus coeruleus of the pons, as well as the medulla and hypothalamus. Noradrenergic projections from the locus coeruleus go to the spinal cord and nearly every major area of the brain, where they increase arousal and vigilance. Norepinephrine at the postganglionic synapses in the sympathetic portion of the peripheral nervous system mediates arousal.

At least four receptor types respond to either norepinephrine or epinephrine. All these receptors are metabotropic. Norepinephrine receptors in the peripheral nervous system mediate fight-or-flight responses.

Below: *Norepinephrine receptors are responsible for determining the fight-or-flight response.*

The *indolamines*: serotonin and melatonin

Indolamines, including serotonin and melatonin, are similar in chemical structure to the catecholamines. About 200,000 serotonergic neurons are in the raphe nuclei of the brainstem. This small number of neurons projects widely throughout the neocortex, limbic system, and cerebellum of the brain, as well as the spinal cord. Serotonergic activity influences behaviors such as sleep, mood, and appetite.

ATP

Adenosine triphosphate (ATP) and its byproducts, particularly adenosine, act as neurotransmitters in the central and peripheral nervous system. Its presence is associated with the perception of pain. ATP often coexists in synaptic vesicles with other neurotransmitters, particularly catecholamines. Adenosine, a breakdown product of ATP in energy metabolism, inhibits the release of several neurotransmitters.

Neuropeptides

Neuropeptides (short proteins) are chains of amino acids. Neuropeptides can also act as *neuromodulators* and *neurohormones* as well as neurotransmitters. There are at least 40 different peptides that act as neurotransmitters, neuromodulators, and neurohormones.

Neuropeptides often coexist in the same neuron with a small-molecule neurotransmitter and modify their effects. A single neuron can contain and release several different neuropeptides.

Among the neuropeptides are substance P, involved in the perception of pain, and the endogenous morphines

Above: *The serotonergic neurons influence several important behaviours, including sleep.*

(endorphins) that act on the same receptors as opiate drugs. There are several peptides associated with digestion, such as insulin and cholecystokinin (CCK) that also have neurotransmitter functions in the brain. Some peptides released from the pituitary gland, such as oxytocin and vasopressin, act as both neurotransmitters and hormones.

Gaseous neurotransmitters

Gases produced within neurons, such as nitric oxide (NO) and carbon monoxide (CO), can diffuse through neural cytoplasm and influence adjacent neurons. These gases, often produced in postsynaptic neurons, diffuse through cytoplasm and membranes without needing vesicles or a release mechanism. They may even travel through one cell and influence its neighbors. Gaseous transmitters often transfer feedback information from the postsynaptic neuron to the presynaptic neuron. They act on receptors located within cells rather than on membrane receptors. Gaseous transmitters break down rapidly by normal chemical reactions and thus are short-lived.

Nitric oxide is involved with neural communication, the maintenance of blood pressure, and penile erection. The anti-impotence medication sildenafil citrate (Viagra) acts by boosting the activity of NO in the penis. NO appears to play a particularly important role in regulating communication between the thalamus and the cerebral cortex, which in turn influences the amount of sensory input processed by the highest levels of the brain. About 2 percent of the neurons scattered throughout the brain release NO.

SENSING THE INTERNAL AND EXTERNAL WORLD

All lifeforms, including single cells like bacteria, sense energy or substances in their environment and generate responses to what is sensed. Animals evolved specialized cells that sense both the external environment (light, sound, touch, odorants, taste…) and the internal environment of their own bodies (temperature, CO_2, blood pressure, glucose…). In the following pages, we will see how the main senses mediate responses to environmental input, namely, vision, audition, skin senses, smell and taste. All of these senses involve receptors that respond to particular physical stimuli and send signals to the thalamus which are then passed on to the relevant area of the cortex. How we experience the world is a unique creation of our brain made possible by these complex sensory processes.

The pathways of the olfactory system—how the sense of smell reaches our brain.

// Vision

Right: *Visual pathways through the brain: the optic nerves reach from the retinas at the back of each eye to the optic chiasma at the brain's center. The optic nerves terminate in the occipital lobes, where visual processing happens.*

When we see the world around us, a series of processes occur that convert the light that enters our eyes into a picture that we can make sense of. Light from objects outside the eye is focused, then captured by photoreceptors whose activity is processed by neural circuitry in the retina and sent to the brain.

To produce an image of an external object, a lens must bend the path of incoming light rays (called refraction) so that all light rays emanating from each point of the object arrive at the same point on the image. The mechanical parts of the eye that are like a camera consist of the cornea, pupil (gap in the iris), and lens.

We can think of the operation of the eye in four main stages:

1. Mechanical focusing and intensity control by the front of the eye.
2. Photon capture by the photoreceptor layer.
3. Processing of photoreceptor signals by retinal circuitry.
4. Transmission of visual information to the brain by retinal ganglion cells.

After light passes through the lens, it passes through a clear, jelly-like substance that fills the entire inside of the eye, called the vitreous, before striking the retina. The retina is a thin neuropil (conglomeration of neurons) that lines nearly the entire inside of the eyeball.

Light passes through all the various retinal cells before reaching the photoreceptors, which are the furthest neurons from the center of the eye. These retinal neurons are highly transparent so that they do not interfere with light reaching the photoreceptor layer.

Photoreceptor outer segments are interlocking in a structure called the pigment epithelium that helps nourish the photoreceptor, supply it with photopigment for absorbing light, and absorb stray light not captured properly by the photoreceptor outer segments.

eyes

NEURAL TYPES IN THE RETINA

- Photoreceptors (rods and cones)—captures light.
- Bipolar cells—modulate changes in light level.
- Horizontal cells—modulate signals from photoreceptors to bipolar cells.
- Ganglion cells—collect visual information from bipolar and amacrine cells.
- Amacrine cells—modulate signals from bipolar to ganglion cells.

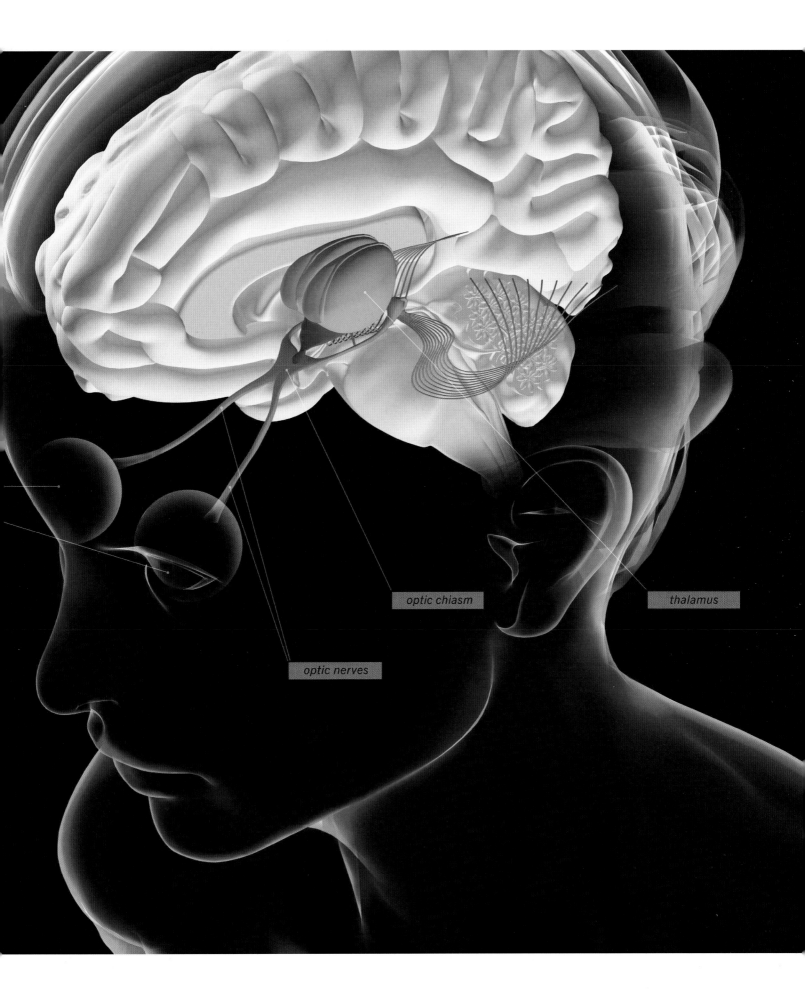

optic chiasm

thalamus

optic nerves

The main information pathway through the retina is from photoreceptors to bipolar cells to ganglion cells (which send their axons to the brain).

The photopigment molecule that captures light is located in the membranes of disks in the photoreceptor outer segment. The inner segment contains photopigment manufacturing machinery, and the cell body. Finally, the photoreceptor terminal, like the axon terminal of any neuron, releases a neurotransmitter when depolarized. However, unlike most axon terminals, this does not depend on action potentials (spikes), but instead occurs continuously in the absence of light.

Photoreceptors

There are two major types of photoreceptors, called rods and cones. Rods function in dim light, such as at night, while cones function in bright daylight. Essentially, rods can be thought of as always turned "on" in the absence of light. The outer segments of both rods and cones that capture light are embedded in the pigment epithelium. These outer segments are densely packed with photopigment.

There are three main sub-structures in rod and cone cells: the outer segment, the inner segment, and the glutamate-releasing terminal. In rods, the visual pigment, rhodopsin, is made up of two molecules—a protein portion called opsin, which does not absorb light, and a light-absorbing portion, called retinal.

Below: *Rods and cones.*

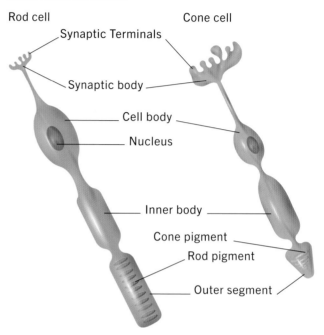

Below: *Structure of the retina.*

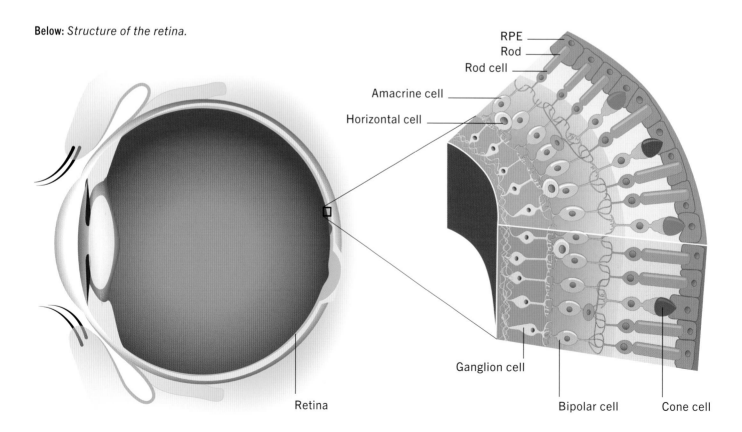

PHOTOTRANSDUCTION

The process by which the light captured by photoreceptors is converted into electrical signals that can be interpreted by the brain is called phototransduction.

Step 1: Light enters the eye and activates pigment molecules in the photoreceptors.

Step 2: The retinal changes shape from 11-cis to all trans.

Step 3: As the retinal no longer fits in the binding site of the opsin molecule, it "pops off" the opsin, and the opsin itself changes its configuration.

Step 4: The opsin conformational change activates a "G-protein," a molecule that causes a messenger molecule, GTP (guanosine triphosphate), to be increased inside the cell.

Step 5: After this, the freed GTP molecules activate another molecule in the disk membrane called phosphodiesterase (PDE). This enzyme converts an "internal neurotransmitter" called cGMP (cyclic Guanosine MonoPhosphate) from its active form to an inactive form, GMP. The function of the active cGMP molecule is to bind to and open sodium ion channels in the photoreceptor membrane, which depolarizes the photoreceptor.

Step 6: With the reduced cGMP concentration, the number of open sodium channels falls, which hyperpolarizes the photoreceptor, reducing the release of glutamate, in essence "turning off" the rod.

Step 7: When the rod "turns off," the bipolar cells it connects to are turned on, sending a signal to the ganglion.

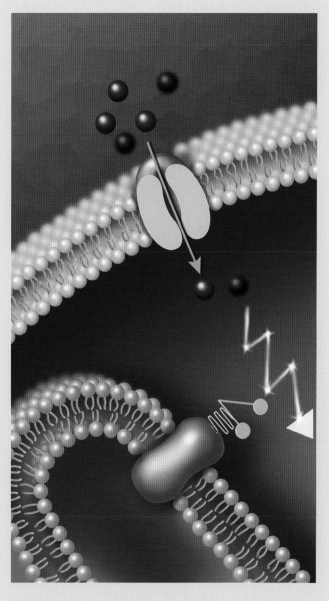

Above: *The phototransduction cascade.*

Bipolar cells

There are two types of bipolar cells—depolarizing and hyperpolarizing. How does the release of glutamate cause opposite responses in these two bipolar cells? The answer is that these two cell types have different receptors for glutamate. When the photoreceptor receives light, it hyperpolarizes, releases less glutamate, and therefore, the hyperpolarizing bipolar cell hyperpolarizes to light. The depolarizing cell has a metabotropic receptor for glutamate. Glutamate binding to this receptor causes potassium channels to open, which are inhibitory, so when the photoreceptors absorb light and hyperpolarize, releasing less glutamate, there is less inhibition of the depolarizing bipolar cells, which depolarize (are excited).

The result of the opposite effect on the two bipolar cell types is a push-pull response to light. At any time, in any scene you are looking at, there is an average light level. As you move your eyes, or things in the scene move, the light level reaching one photoreceptor moves up or down compared to this average light level. Depolarizing bipolar cells signal positive light level changes with excitation, while hyperpolarizing bipolar cells signal negative light level changes.

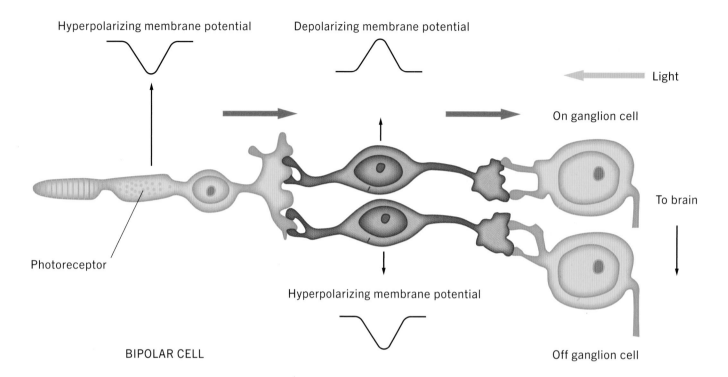

Hyperpolarizing membrane potential

Depolarizing membrane potential

Light

On ganglion cell

To brain

Photoreceptor

Hyperpolarizing membrane potential

BIPOLAR CELL

Off ganglion cell

Horizontal cells

The other cell type that receives glutamate from photoreceptors is the horizontal cell. Virtually all photoreceptor terminals have multiple synaptic terminal complexes, in which there is a central bipolar cell dendrite, flanked on each side by a dendrite of a horizontal cell.

The horizontal cell dendrites have AMPA/Kainate receptors for glutamate (see page 68), so they hyperpolarize to light, like the photoreceptors. The function of horizontal cells is to inhibit bipolar cells. The horizontal cells receive inputs from photoreceptors surrounding the central photoreceptor. Via synaptic gap-junction coupling (see page 61) among the horizontal cells, the horizontal cell population activity represents an estimate of the average light level. This is subtracted, via inhibition, from the output of the central photoreceptor. The result is that the bipolar cell responds to the difference between the photoreceptor(s) to which it is directly connected, and the surrounding ones. This is called contrast. The visual system codes the presence of objects and object borders via changes in illumination level, not the absolute illumination level.

Ganglion cells

Bipolar cells send their outputs to amacrine and ganglion cells. Bipolar cells also release the neurotransmitter

glutamate. The ganglion cells that receive input from depolarizing bipolar cells will receive more glutamate when stimulated by light above the average level, firing more action potentials. These are called On-center ganglion cells, meaning they fire more action potentials to light. Ganglion cells that receive input from hyperpolarizing bipolar cells will fire more action potentials for dark areas of the scene. These are called Off-center ganglion cells.

The firing of action potentials that are shipped out from the ganglion cell axons to various brain structures inches away reveals the reason for the push-pull depolarizing/hyperpolarizing bipolar cell arrangement. Because of the metabolic expense of making action potentials, most ganglion cells (and neurons in general) maintain only a low level of maintained or "spontaneous" firing. On-center ganglion cells increase firing for light increases. But, if they are not firing spontaneously, they cannot signal much of a range of light decreases, because they cannot fire fewer than zero spikes. Hyperpolarizing bipolar cells, and the Off-center ganglion cells to which they are connected, signal light decreases in a parallel way.

Amacrine cells and complex ganglion cell classes

Depolarizing and hyperpolarizing bipolar cells also send outputs to amacrine cells. Amacrine cells contact each other as well as ganglion cells. One of their functions is to mediate the same sort of inhibition for the connections between bipolar and ganglion cells, as horizontal cells do for connections from photoreceptors to bipolar cells.

Amacrine cells are much more complex than horizontal cells, however. Whereas there are only two major classes of horizontal cells, there are at least 50 classes of amacrine cells. What most of these amacrine cell classes do is sculpt the output of bipolar cells to produce ganglion cells selective for specific patterns of light distribution. For example, some ganglion cells respond only to the presence of edges, others only to certain colors, others only to movement in a certain direction.

Retinal ganglion cell axons project to at least 20 different structures in the brain that perform a variety of visual functions, ranging from control of pupil diameter, to compensating for head movements to maintain a stable image on the retina, as well as, of course, conscious perception of objects in the world.

The retina has layers with various cell types. Furthest from the center is the outer nuclear layer that contains the inner segments of photoreceptors. Next to that is the outer plexiform (synaptic) layer, which contains the photoreceptor terminals and dendrites of bipolar and horizontal cells. The inner nuclear layer contains the cell bodies of horizontal cells, bipolar cells, and amacrine cells. The inner plexiform layer contains the axon terminals of bipolar cells, and the dendrites of amacrine and ganglion cells. Last, the ganglion cell layer contains the cell bodies of ganglion cells.

There appear to be over 30 classes of ganglion cells in the mammalian retina. Two ganglion cell classes crucial for visual perception that project to the visual area of the thalamus are called brisk-sustained and brisk-transient in some mammals, X and Y in others, and parvocellular and magnocellular (see pages 80–81) in humans.

Below: *Microscope image of the retina.*

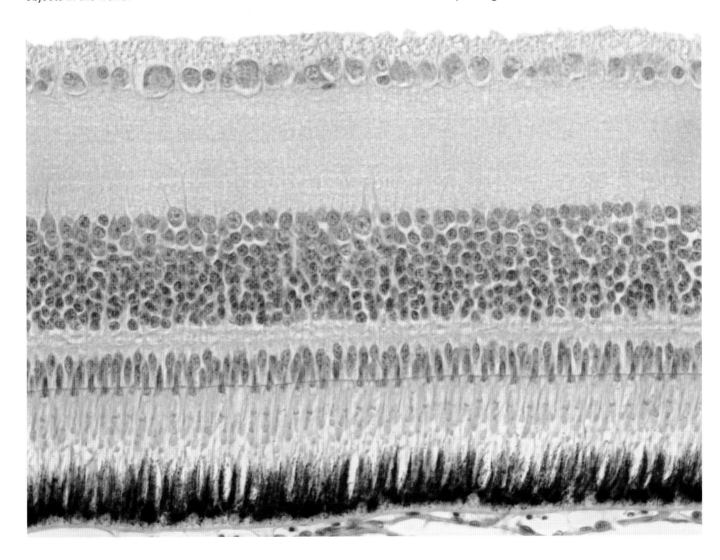

// Visual pathways

Visual processing proceeds from the capture of photons by photoreceptors to the activation of bipolar cells to ganglion cells. Lateral interactions in the retina are mediated by horizontal cells and amacrine cells.

The optic nerve

The axons of ganglion cells project to brain targets many centimeters from the retina, requiring them to send action potentials because of the distance. The bundle of all retinal ganglion cell axons leaving the eye is called the optic nerve. The human optic nerve contains the axons of slightly over one million ganglion cells. Within a few centimeters after the left and right optic nerves leave the two eyes, they meet in a structure called the optic chiasm. Here the axons from the two eyes are rearranged in a specific manner to form what are called the two optic tracts. Ganglion cell axons from the temporal retina (the part of the retina farthest from the nose) of each eye project to the same left or right side as the eye of origin, while ganglion cells from the nasal retina (the part closest to the nose) project to the opposite optic tract. This means that the right optic tract contains axons from both

Right visual field | Binocular field | Left visual field

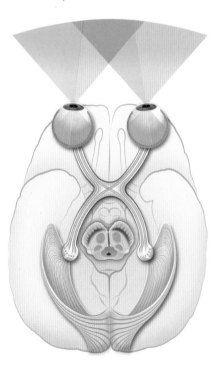

Above: *At the optic chiasm, the optic nerves go from the left visual field of both eyes to the right hemisphere of the brain and vice versa.*

eyes whose ganglion cells are stimulated by the left visual field, with the reverse being true for the left optic tract.

Brain targets of the optic nerve

The output of the eye allows us to do more than consciously see objects in the world. The outputs of some retinal ganglion cells go to brain areas that project back to the eye and control lens accommodation (power) and pupil diameter. Some ganglion cells mediate circadian rhythms that control the day/night cycle. Yet other ganglion cells detect head motion for control of balance and image stabilization. There are at least 30 classes of retinal ganglion cells in mammals whose response properties are appropriate for these different tasks. These different ganglion cell classes project to at least 15 different brain targets. Understanding the output of the eye entails understanding the information each ganglion cell class extracts from the visual scene, and what brain circuits use that information.

The most important brain targets of the optic nerve are:

1. The lateral geniculate nucleus (LGN) of the *thalamus*.
2. The *superior colliculus* in the midbrain.
3. The *accessory optic system*, consisting of several nuclei in the brainstem.
4. Nuclei called *pretectal* nuclei in the midbrain.
5. The *suprachiasmatic nucleus* in the hypothalamus.
Most (though not all) retinal ganglion cell classes project to several targets.

The thalamus

Conscious visual perception depends on the projection to the thalamus. The projection to the lateral geniculate nucleus of the thalamus, which then projects to visual cortex, mediates our conscious perception of the visual world. If this pathway is severely damaged, some visual functions, such as pupil reflexes and circadian rhythms, may remain intact, but there will be no conscious perception of visual input.

Two major classes of retinal ganglion cells constitute the main projection to the lateral geniculate nucleus (LGN) of the thalamus:

1. Parvocellular ganglion cells that mediate pattern and high acuity vision.
2. Magnocellular ganglion cells that mediate detection of change and motion.

Parvocellular ganglion cells are responsible for high acuity and color vision. They have small receptive fields, and some have color discrimination. Magnocellular ganglion cells have larger

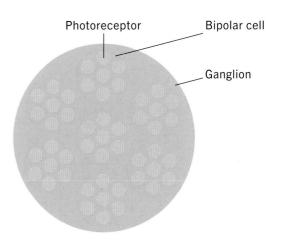

Above: *Several photoreceptors (red) synapse on a single bipolar cell (green). Several bipolar cells synapse on a single ganglion cell (blue). In this case, one ganglion cell is relaying activity from a group of 49 photoreceptors. When many precursor-neurons synapse on a single output neuron like this it is called* convergence.

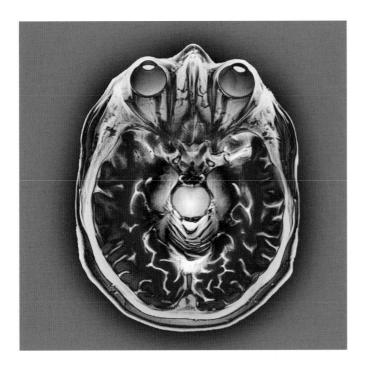

Above: *An MRI scan of a human brain and eyes.*

receptive fields and respond well to movement and change but are poorer at signaling exactly where the change is located.

The dendritic coverage of each ganglion class (on or off-center) is highly organized, with no gaps and little overlap (called "tiling"). The area of the retina where changes in distribution of incoming light modulates the responses of a ganglion cell is generally about equal to its dendritic tree extent.

One implication of this is that each point on the retina is in the receptive field of exactly one ganglion cell of each class. In other words, from each point in the retina, each of the 30 ganglion cells extract some unique feature of the visual scene to send to the brain. Each large, magnocellular receptive field is covered by eight–ten smaller parvocellular receptive fields. As ganglion cells respond similarly to changes in the light distribution anywhere within their receptive field, they cannot signal where within the receptive field the light change occurred. Position acuity occurs when moving a point of light across the retina causes it to activate a different cell at the second, rather than at the first position. High positional acuity requires, therefore, small receptive fields (and lots of them).

Different ganglion cell classes that have different functions in vision are of different receptive field sizes. However, all ganglion cell classes increase in size as a function of distance from the center of the retina (called *eccentricity*).

The density of ganglion cells is much higher in the center than the periphery. The one point where this density function is violated is called the blind spot, where the axons of all retinal ganglion cells begin the optic nerve and leave the retina. There are no ganglion cell bodies here (or photoreceptors, for that matter).

The parvo- and magnocellular ganglion cell classes project to different layers of the lateral geniculate nucleus of the thalamus (LGN), of which there are six layers in primates. The total number of neurons in all these layers is very nearly the same as the number of ganglion cell inputs. Most LGN cells are driven primarily by a single retinal ganglion cell.

Neurons in each LGN that receive retinal input project, in turn, to the primary visual cortex in the occipital lobe via fiber tracts called optic radiations. All sensory systems, including vision (but with a partial exception for olfaction), project to a specific area or nucleus of the thalamus, which in turn projects to a specific area of the neocortex.

PRIMATES COMPARED TO OTHER MAMMALS

The retinas of non-primate mammals typically contain about 100,000 to 200,000 ganglion cells—mammals with large eyes having somewhat more than those with small eyes. Typically, the ganglion cell density is higher in the center of these retinas, called the *area centralis*, than in the periphery. What primates do is add many ganglion cells in the very center of the area centralis, the *fovea*, a small area of ultra-dense neurons. In humans this is taken to an extreme, where the foveal region contains about one million ganglion cells. Thus, in humans, and primates to a lesser extent, most of the output of the eye is from the small foveal region.

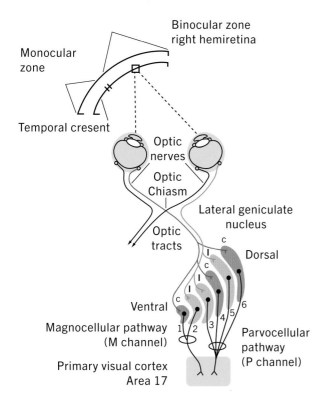

Above: *Ganglion cell projections to LGN layers.*

The cortical areas that receive direct input from the thalamus are called V1, for vision, A1, for audition, S1, for somatosensation (skin senses), and so forth. These primary areas then project to "higher" or secondary areas of cortex, which may in turn project to even higher areas.

The projection from retina to LGN is about one to one, with each retinal ganglion cell driving one LGN cell with similar response properties. This changes greatly when V1 is reached. There, each LGN cell axon drives 100–200 cortical neurons. The cortical neurons in V1 are more selective in their response properties, firing spikes only when, for example, an edge is present of a particular orientation, movement direction, and length.

The three major classes of V1 neurons are:

1.   Simple—orientation selective.
2.   Complex—orientation and direction of movement selective.
3.   Hyper-complex or end-stopped—orientation, direction of movement, and length selective.

An important aspect of this response selectivity is that the representation of information from the visual scene is embodied in a small percentage of all the neurons at that cortical level. This is called a sparse representation. Few neurons fire for any given visual pattern, but the firing of those few neurons represents a lot of information about the visual pattern on that part of the retina. As one moves from V1 to "higher" cortical areas, the response selectivity becomes more specific.

Major visual cortical areas

The primary visual area V1 projects to the area V2, where cells have similar response properties. After the V1–V2 complex, however, visual cortical processing splits into two primary streams, called ventral and dorsal.

Dorsal stream

The dorsal stream (so named because it involves areas toward the top, or dorsal surface of the brain) proceeds to area V3, then an area called MT (also called V5), and then to several areas of the parietal lobe. These parietal lobe areas project to motor areas of the frontal lobe to provide the basis for visually guided behavior. Thus, the dorsal stream visual system is what you use to catch a baseball or run through the woods without colliding with the trees. Neurons in this pathway are often sensitive to motion or motion "flow fields" created by self-movement.

Ventral stream

The ventral stream involves a projection from V2 to area V4, and then to a succession of cortical visual areas in the inferior temporal (IT) lobe. Neurons in these areas respond to complex visual patterns and their activity underlies our ability to recognize and perceive objects. The IT lobe is heavily interconnected with the hippocampus because perception and recognition depend on memory.

Below: *Dorsal and temporal processing visual streams.*

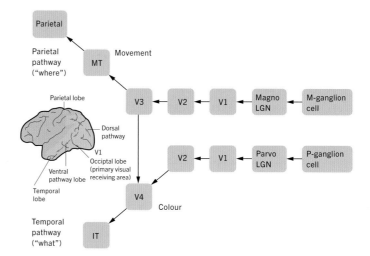

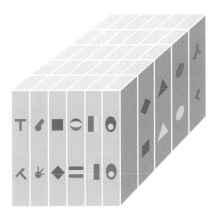

Above: *Visual pattern selectivity in IT cortex.*

The superior colliculus

Another target of retinal ganglion cells is the *superior colliculus* in the midbrain (part of the brainstem). Most ganglion cell classes, except for parvocellular ganglion cells, project to the superior colliculus. Non-mammalian vertebrates, such as lizards and frogs, have a structure called the tectum that is the homolog of the mammalian superior colliculus. Activation of the tectum by ganglion cell input in these animals causes reflexive behavior.

In mammals the superior colliculus controls one type of the eye movements that occur when we scan a scene or read a page of text, moving our eyes from point to point. These movements are called *saccades*, with the holding of gaze between saccades called *fixations*. Primates typically make about three or four saccades per second. We are generally not consciously aware of the targets for these movements. The output of virtually all ganglion cell classes, except parvocellular cells, sets up the computation for these saccades.

The superior colliculus projects to a structure in the thalamus called the pulvinar, which in turn projects to the parietal lobe, which in turn projects to the frontal lobe areas that control eye movements. The pulvinar also receives from and projects to high order visual areas in the dorsal stream to control visual attention, as such when one is voluntarily tracking a moving object.

The accessory optic system

The accessory optic system is comprised of three major brainstem nuclei called the dorsal, medial, and lateral terminal nuclei, located in the brainstem. These nuclei are named "accessory." A large percentage of the ganglion cells that project to these nuclei respond specifically only to a particular direction of motion. Two of the important functions of these nuclei are:

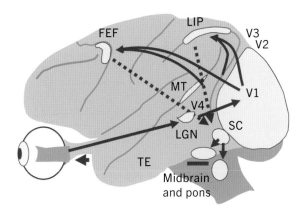

Above: *Visual pathways for saccades.*

1. Providing a parallel input to that of the vestibular system that signals head (and body) movement for balance.
2. Signaling "slip" when tracking objects with the eyes to keep the tracked object stationary on the fovea.

Early in life, when the visual system is relatively undeveloped, balance depends mostly on the output of the semicircular canals in the inner ear to indicate head movement and rotation. Once the visual system develops, however, it is more accurate than the vestibular system and projects to the same nuclei where it may override the vestibular input. The two systems can cause problems when in conflict, however, as when, after spinning around, or in micro-gravity, nausea may occur.

Pretectal nuclei

The pretectal nuclei at the upper border of the midbrain receive retinal input that allows them to control functions like pupil diameter and smooth pursuit eye movements (the latter in conjunction with the accessory optic nuclei). Some researchers believe some pretectal nuclei are important controllers of the initiation of REM (rapid eye movement) sleep.

Suprachiasmatic nucleus

The suprachiasmatic nucleus is a nucleus of the hypothalamus located just above the optic chiasm, as its name implies. This nucleus receives inputs from several classes of what are called *intrinsically photosensitive* retinal ganglion cells. These ganglion cells have light absorbing protein complexes in their dendritic membranes that cause them to directly respond to light by modulating their firing rate of action potentials. Photosensitive retinal ganglion cells that project to the suprachiasmatic nucleus regulate circadian rhythms by resetting the circadian clock to sunrise or light onset each day.

// Visual dysfunction

The visual system begins with the front of the eye performing a camera-like focusing of an image of the external world on the retina. As we touched on earlier, after retinal processing, at least 30 classes of retinal ganglion cells project visual information to over 15 retinal recipient zones in the brain. Neurons in almost half of the neocortex, comprising over 30 areas, can be driven by visual input. Visual dysfunctions can arise at any stage in this sequence. Visual problems range from optical blurring of the image due to front of the eye problems, to subtle perceptual agnosias (loss of some specific visual perception capability, such as the ability to recognize faces) from cortical damage, to blindness.

While many issues in visual perception stem from damage to the eye itself—ranging from defects in the shape of the cornea causing image blurring (astigmatism) to opacities in the lens (cataracts)—some issues arise further down the chain, from the operation of photoreceptors in the retina to issues in the optic nerve and in the brain itself.

Most retinal visual dysfunction involves, at least initially, photoreceptor damage, with the important exception of glaucoma, a disease of retinal ganglion cells.

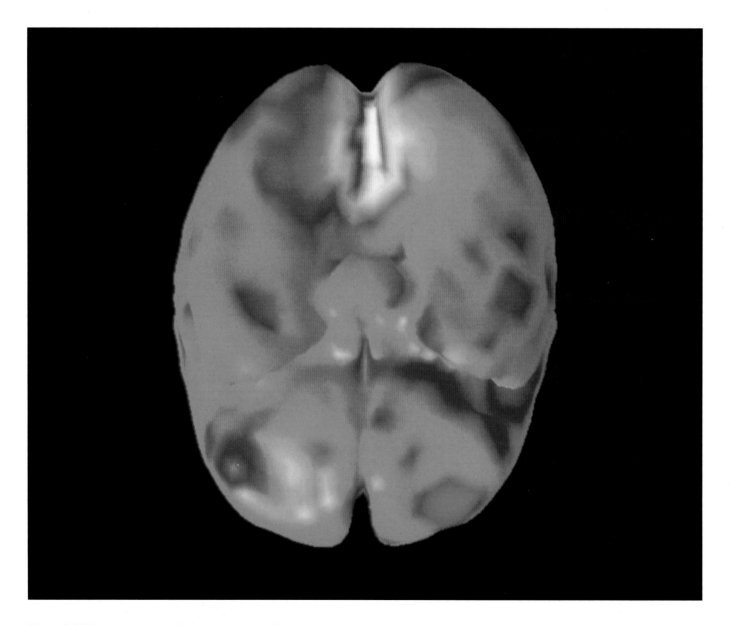

Above: *A PET scan of the brain during visual activity.*

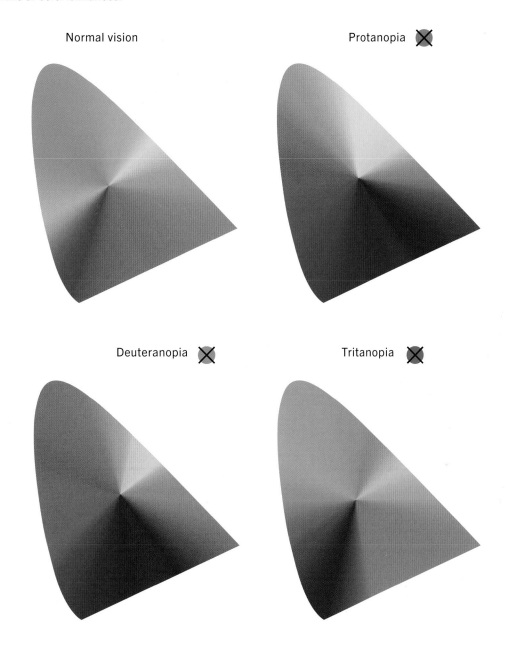

Normal vision

Protanopia

Deuteranopia

Tritanopia

Color blindness

Most of us know someone who is "color blind," usually with the inability to distinguish red from green. There are three major types of color blindness, called:

1. *Protanopia*—loss of red cone function.
2. *Deuteranopia*—loss of green cone function.
3. *Tritanopia*—loss of blue cone function.

Normal color vision based on three cones is called *trichromatic*, with the loss of any one cone causing the person to become what is called a *dichromat*. In protanopia,

a mutation causes the pigment in the red cones to have a similar absorption spectrum to the green cones. In deuteranopia, the pigment in the green cones becomes, by a mutation, red sensitive.

The genes for the red and green cone pigments are on the portion of the X chromosome that is not matched by a homologous sequence on the Y chromosome, making red/green dichromatism a sex-linked trait, being more common in males as males have only one X chromosome. The incidence of male red-green color blindness is about 1/20. Because females have two X chromosomes, however, the non-mutated X chromosome can manufacture enough of the correct red

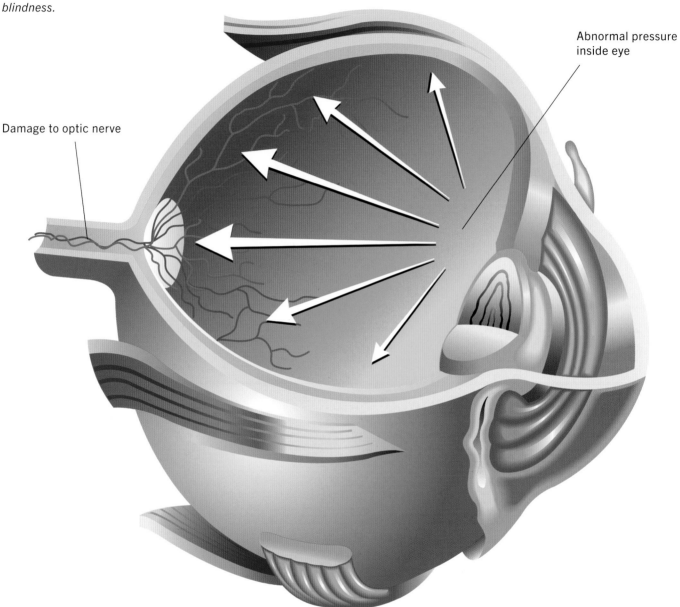

Abnormal pressure inside eye

Damage to optic nerve

or green pigment to prevent them from being red-green color blind. For females to be red-green colorblind, they must inherit mutations on both their X chromosomes, so the incidence of this condition in females is the square of that in males, or 1/400.

Tritanopia is rarer and has a different cause. Tritanopia occurs not from a switch from blue-sensitive to red or green, but from a loss of blue cones, which are much rarer than red and green cones in the normal eye.

Blindness

Much rarer than dichromatism is the condition where all cone function is absent, called *rod monochromatism*. This is caused by mutations that cause the death of all cones early in development. People with this condition cannot see in any but dim light and have very low visual acuity even in very dim light.

1. Loss of rod function

There are several congenital and metabolic abnormalities that compromise rod function. In many of these cases, although loss of rod function is the initial visual deficiency, the death of rods brings about a cascade of events that leads to the death of cones also and to blindness. There is little regeneration capability of the cells in the retina, so that, if any population of cells like photoreceptors is lost, the effect is permanent.

Retinitis pigmentosa is a congenital condition that begins with the degeneration of rods in early adulthood, at first causing night blindness, but later leading to cone death starting in the periphery and then progressing toward the center of vision. Prior to loss of central vision, the patient experiences what is called "tunnel" vision, like looking through a tube, where sight exists only in the very center of the visual field.

Glaucoma, caused by the death of retinal ganglion cells due to excessive pressure within the eye, is a leading cause of irreversible blindness.

2. Visual deficiencies originating after the eye

Damage to the visual system anywhere beyond the eye will cause specific vision problems depending on location. Such damage can be the result of penetrating brain injury, tumors, vasculature accidents such as strokes, and other factors. Damage to the optic nerve, such as with glaucoma, will cause loss of vision in the eye of origin of that nerve. Damage to the optic tract will cause loss of vision in the left or right visual field in both eyes.

3. Amblyopia

Amblyopia is the loss of visual function in one eye even though the retinal output of that eye may be normal. Typically, amblyopia occurs because of a severe congenital optical problem in one eye, such as a cataract or severe focusing problem. During normal development there is a period in which the projections from the two eyes compete for synapses on cortical neurons in primary visual cortex (V1). If one eye has a severe optical problem, the ganglion cells in that eye will have reduced output, and the LGN axons from the more normal eye, with their higher activity, will take over most of the cortical synapses.

The critical period for this developmental neural wiring is the first six years in humans. If the optical problem in the "bad" eye is not corrected during this time, the plasticity period ends. At this point, even if the optical developmental problem is corrected so that the output of the bad eye is normal, it will never be able to drive visual cortical neurons, and the person is paradoxically blind in that eye even though the eye itself may appear to be normal.

4. Damage to the lateral geniculate nucleus of the thalamus

Damage to the thalamus on one side of the brain will, like damage to the optic tract, cause a loss of visual function in the visual field on the opposite side. This is also true for damage to the optic radiation, the axonal pathway of projection from the thalamus to visual cortex.

5. Damage to the visual cortex

Damage to the ventral stream tends to compromise the perception of fine detail, color, and the ability to identify objects. Damage to the dorsal stream affects the ability to use vision to guide movement, and then compromises the ability to detect movement itself.

6. Ventral stream damage

Damage to neurons or neural tracts along the ventral stream produces increasingly subtle perceptual losses as a function of how far the damage is from the V1/V2 complex. A major early station in the ventral stream is area V4. Damage to this area produces loss of the ability to discriminate fine detail and a specific color deficiency called *achromatopsia*.

Achromatopsia is a deficit in the ability to perceive color, without the loss of the ability to see the difference between areas of different color. To someone with achromatopsia, different colors appear as different "dirty shades of gray." This differs from retinal color blindness, where no difference at all can be seen between two areas of different color solely based on color.

On the ventral side of the inferotemporal cortex is an area called the fusiform face area on each side of the brain. Damage to either of these areas, but particularly the right fusiform face area, results in the specific inability to recognize faces, but without the loss of most other visual perceptual capabilities, including the recognition of other objects such as tools, or even animals. This is called prosopagnosia.

Damage to more posterior areas of the inferotemporal lobe may result in the loss of the ability to distinguish other perceptual categories (agnosias), such as the inability to distinguish and recognize types of fruit, animals, or common objects.

7. Dorsal stream damage

Damage in the dorsal stream, particularly in the right parietal lobe, results in various forms of what is called neglect, the inability to notice objects in the left visual field. Also compromised is the ability to perceive and manipulate spatial patterns, such as the ability to arrange colored tiles to duplicate a pattern shown on a piece of paper.

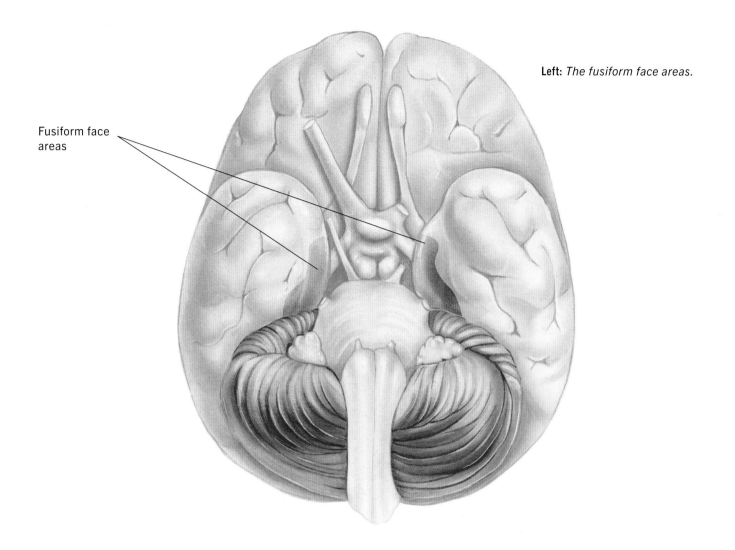

Fusiform face
areas

Left: *The fusiform face areas.*

Hemineglect refers to the lack of awareness of items in half the visual field. It most commonly occurs with damage to the right parietal lobe resulting in neglect of the left visual field. In severe cases, a patient may only shave the right side of his face, for example, and be unaware of totally missing the left side when looking in the mirror. In less severe cases, patients may notice items in the left visual field if there is nothing in the right visual field but will ignore the left visual field if there is any item of interest on the right side. Damage to the left parietal cortex produces much less severe neglect deficiencies, indicating a processing specialization for visual items on the right side of the brain.

Blindsight
The projection of retinal ganglion cells to over 15 retinal recipient zones in the brain enables visually guided behaviors ranging from perception to control of eye movement, to circadian rhythms. Conscious perception depends on the projection pathway to the LGN and then to V1 in cortex.

Patients who have suffered extensive damage to V1 are blind in the corresponding areas of the visual field, reporting no awareness of any stimuli presented in that area. Yet, paradoxically, these patients, when asked to make an eye movement to location of stimuli in their blind visual fields, can do so quite accurately. This is called *blindsight*, the ability to use visual input of which the person is unconscious to guide behavior.

Researchers suggest that blindsight is based on the projection to the superior colliculus that normally controls eye movements without conscious direction (see page 83). For example, when we read, we make nearly optimal eye movements from fixation point to fixation point without any explicit awareness of the content of the text at the new position.

Opposite: *The causes for the loss of visual function can be located anywhere along the optic pathway, from the photoreceptors in the retina to the brain itself.*

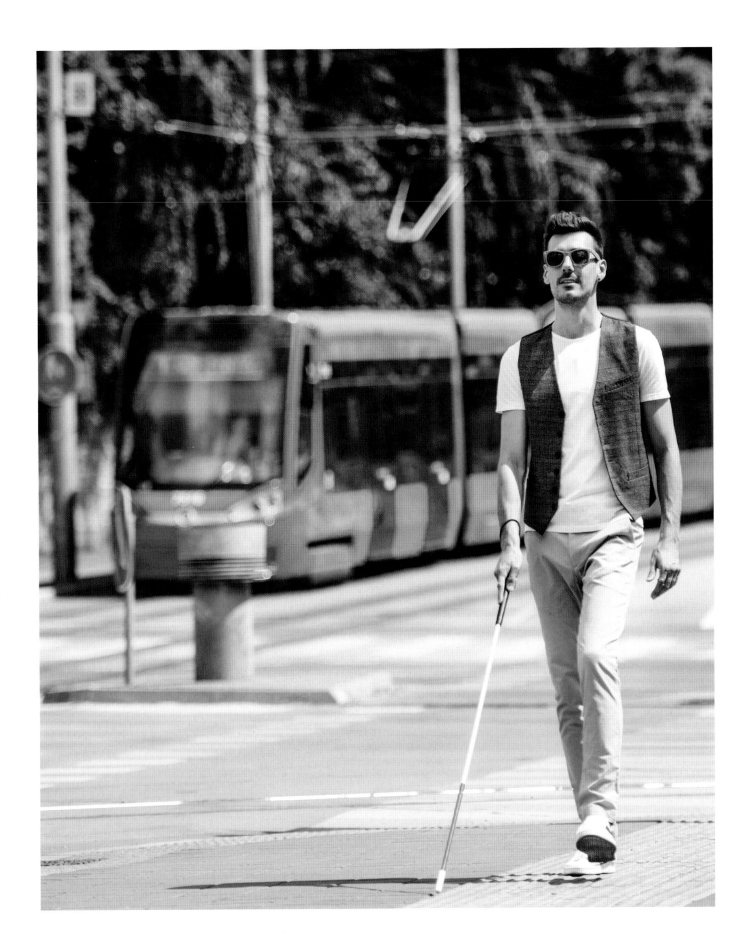

// The auditory system

Vibrating objects, such as a loudspeaker diaphragm, produce sound when they compress air on the side the diaphragm is moving toward, and rarify air on the opposite side. The difference between the maximum compression and rarefication is called the amplitude or intensity. We experience this physical intensity as loudness. Pure tones occur when an object vibrates at a single frequency. The number of cycles per second of the physical frequency is perceived as pitch. Our ears respond to a million-fold range of sound intensities, conventionally expressed in logarithmic units called decibels.

Below: *The auditory system.*

Hearing involves the capture and transduction by the ears of sound waves that have propagated away from the source that emits them. As with the eye, there are structures in the outer parts of the ear that perform "mechanical" transformations of the sound energy. After this, sound transducer neurons in the inner ear begin the process of generating a neural representation/extraction of information in the sound wave.

The path of sound

The ear apparatus is divided into the outer ear, middle ear, and inner ear. The outer, or external ear, consists of the *pinna*, *auditory canal* (external acoustic canal), and tympanic membrane (*eardrum*). The pinna (colloquially called the "ear") reflects sound waves into the auditory canal. The auditory canal conveys sound to the eardrum. Sound energy

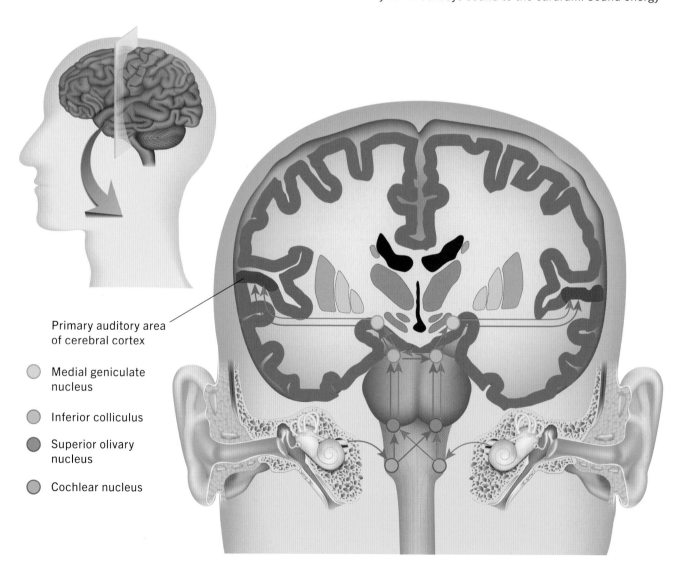

Primary auditory area
of cerebral cortex

○ Medial geniculate
nucleus

○ Inferior colliculus

● Superior olivary
nucleus

○ Cochlear nucleus

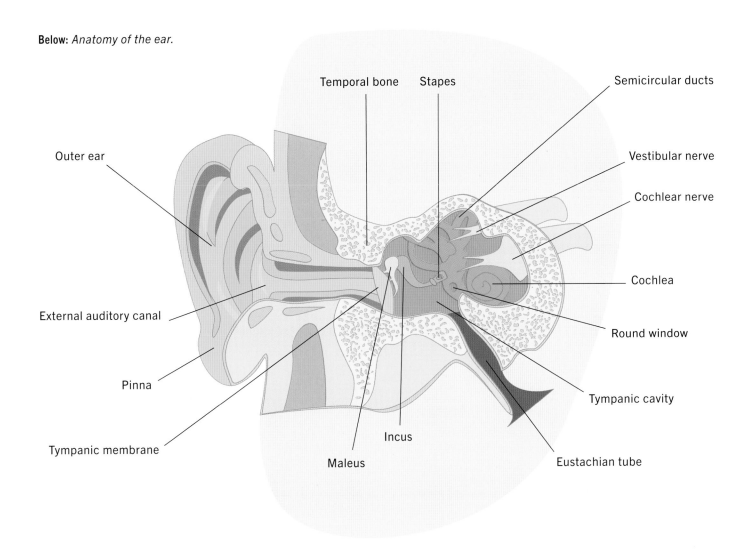

Temporal bone   Stapes

Semicircular ducts

Outer ear

Vestibular nerve

Cochlear nerve

Cochlea

External auditory canal

Round window

Pinna

Tympanic cavity

Incus

Tympanic membrane

Maleus

Eustachian tube

traversing the auditory canal sets the eardrum in motion. Eardrum motion is transmitted to the middle ear that in turn transmits vibrations to the cochlea at the oval window. The middle ear consists of the three bones called the malleus, incus, and stapes, which transmit vibrations from the eardrum to the cochlea at the oval window. One function of these bones is force amplification via leverage.

Vibration of the oval window by the stapes sends sound waves into the fluid filled cochlea, the neural transducing apparatus of the inner ear. In and out movement of the stapes at the oval window causes fluid to move within the cochlear chambers with corresponding passive relaxation movement of the round window. The basilar membrane in the cochlea contains the Organ of Corti, where sound is transduced into neural activity.

Vibrations sent into the cochlea by the stapes at the oval window cause the tectorial membrane to move laterally with respect to the basilar membrane, bending the cilia of the auditory hair cells. There are two varieties of auditory hair

cells, called inner and outer. Most of the auditory information sent to the brain is from the inner hair cells. The more numerous outer hair cells function to control the stiffness of the tectorial membrane, enhancing inner hair cell output at low sound intensity but decreasing it at higher intensity.

The shearing action between the basilar and tectorial membranes deflects the hair cell cilia, which depolarize when bent in one direction, and hyperpolarize when bent in the other. The depolarization phase is driven by the opening of relatively non-selective ion channels allowing the entry of potassium, sodium, and calcium into the hair cell cilia. The extracellular fluid outside hair cells has an abnormally high potassium concentration, so that opening these channels causes a net depolarization of the hair cell that releases the neurotransmitter glutamate.

Glutamate released by the hair cells elicits action potentials in the endings of the spiral ganglia axon-like myelinated neuronal processes. The cell bodies of these processes are in the spiral ganglia. Hair cell cilia deflections as small as the

Below: *The organ of corti.*

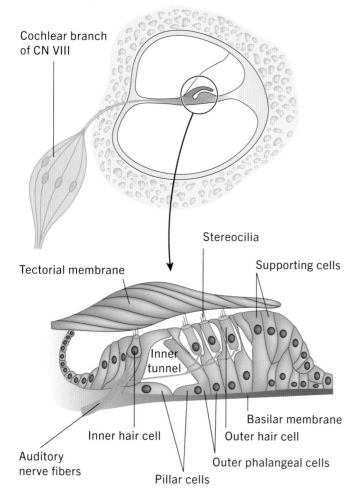

Cochlear branch of CN VIII

Stereocilia

Supporting cells

Tectorial membrane

Inner tunnel

Inner hair cell

Outer hair cell

Basilar membrane

Auditory nerve fibers

Outer phalangeal cells

Pillar cells

AUDITORY NERVE FIBERS

Auditory nerve fibers are frequency selective. The resonant frequency of the cochlea varies along its length due to mechanical properties, such that high frequencies only produce large vibrations near the entrance, near the oval window, while lower frequencies vibrate more distant portions toward the apex (helicotrema). This frequency selectivity along the cochlea is called a "place code." Place coding exists because neurons cannot fire more than 500–1,000 spikes per second, but sound frequency detection in humans covers a range of 20–20,000 Hz. At frequencies below 500 Hz, neurons in virtually the entire cochlea fire at the peak of every sound wave. At higher frequencies, however, auditory neurons can only fire every other sound wave peak, or every third peak, and so forth. Higher frequencies are then encoded by which neurons along the cochlea are firing.

width of an atom can produce action potentials.

About 3,500 inner and 12,000 outer hair cells contact the axons of about 30,000 spiral ganglion cell axons. Most inner hair cells contact more than one spiral ganglion cell axon. Individual auditory nerve fibers are sensitive to particular frequency ranges, especially for higher frequencies, reflecting the frequency map of their position in the basilar membrane (see box).

The cell bodies of the auditory nerve fibers are in a structure called the spiral ganglion. One axonal process in the spiral ganglia projects to the cochlea and contacts auditory hair cells. At its end, it has glutamate receptors that cause a spike to be initiated. This spike proceeds toward the cell body, then continues past the cell body toward the cochlear nucleus in the medulla.

These bifurcations are joined by axons that project to the brainstem. This combined nerve is called cranial nerve VIII. The auditory portion of cranial nerve VIII projects to the cochlear nucleus via the other extension of the spiral ganglion axon. This structure has similarities to that of the dorsal root ganglion cells outside the spinal cord.

The cochlear nuclei, each of which receives input from the spiral ganglia on that same side of the brain, is located in the dorso-lateral brainstem in the medulla close to the border with the pons. This nucleus functions primarily as a relay nucleus, projecting auditory input to the superior olive.

The superior olives on each side of the brain receive input from the cochlear nuclei of both ears. A major function of the superior olivary nuclei is the detection of the horizontal angle of sound direction by comparing the intensity of the sound in the two ears.

The superior olives project to the inferior colliculi in the midbrain. The inferior colliculi are involved in frequency discrimination and signal integration with the superior colliculi.

Neurons in the inferior colliculus project upwards to the medial geniculate nucleus (MGN) of the thalamus. This nucleus projects to auditory cortex in the temporal lobe. Neurons in the ventral area of the MGN are organized as a tonotopic (*frequency*) map, like that of the primary auditory cortex. Neurons in the dorsal part of the MGN respond to complex environmental stimuli, while the medial MGN has neurons that code for horizontal direction (azimuth).

The medial geniculate nucleus projects to the primary auditory cortex (A1) located in the medial area of the temporal lobe. The gyrus at this area is called Heschl's gyrus. The primary auditory cortex is organized as a tonotopic map in a manner similar to that of the basilar membrane in the cochlea.

Neurons in the primary auditory cortex project to other "association" cortical areas for higher order sound

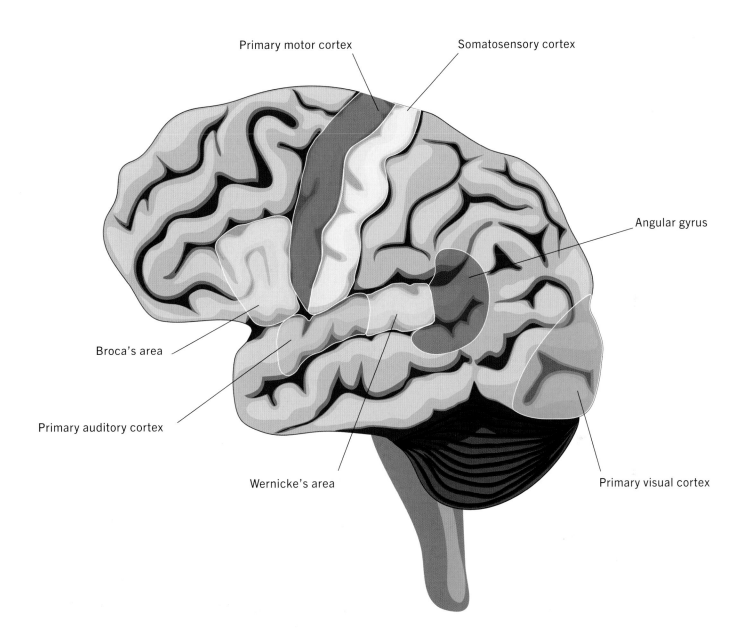

Primary motor cortex

Somatosensory cortex

Angular gyrus

Broca's area

Primary auditory cortex

Wernicke's area

Primary visual cortex

processing. Secondary and higher order ("association") auditory areas surround the primary auditory cortex. These higher order areas have neurons that respond selectively to complex environmental sounds to mediate sound identification.

Hearing loss

When sound is not properly transmitted from the eardrum to the cochlea it is called "conductive" hearing loss, indicating a deficiency in the mechanical transmission of sound prior to neural transduction. This can occur from damage to the eardrum itself (such as rupture from diving), or from damage to the middle ear bones.

Neural hearing loss occurs when transduction or transmission neurons for sound are damaged. The most common form of damage is the loss of auditory hair cells in the cochlea. This can occur gradually during aging (called presbycusis), or from a single exposure to a very loud sound. Hearing loss can also occur due to neural loss at higher levels in the auditory system, from cranial nerve VIII to auditory cortex, from mechanical injury, tumors, or strokes.

// Touch

The skin encloses the body, protecting the inside from dirt, bacteria, and other assaults from outside. However, the skin is far more than just an enclosing boundary because it also detects contact with the world through touch. The touch senses include mechanical displacement, temperature, and pain. Together, these senses are known as *somatosensation*. Each of these different senses is converted into an electrical signal (transduced) by a different type of skin receptor.

Specialized sensory neurons located in the dorsal root ganglia outside the spinal cord are responsible for transducing these senses, while skin senses in the head are mediated by cranial nerves.

The dorsal root ganglion cell has no dendrites or synapses. Instead, the cell body gives rise to a single axon that immediately bifurcates. One end of the axon goes in an efferent direction (toward the periphery). The axon ending contains several different receptor types. These include not only the various skin senses mentioned above, but also receptors for muscle tension, stretch, limb position, and velocity. These latter senses are called proprioception and kinesthesis.

What all these sensory neurons have in common is that their axon endings produce action potentials. So-called "touch" or mechanoreceptors in the skin respond to various kinds of skin deformation. Temperature receptors, as their name implies, respond to specific temperature ranges. Pain receptors, called nociceptors, respond to impending damage to the skin from various modalities ranging from high heat or cold, to chemical assaults, to mechanical tearing.

Action potentials from these sensory neurons travel in a "retrograde" manner toward the axon bifurcation near the cell body in the dorsal root ganglion. These action potentials then continue along the other portion of the axon bifurcation into the spinal cord. The ends of the bifurcating axons enter the spinal cord at the dorsal root of the spinal cord gray area. This axon ending releases the neurotransmitter glutamate at synapses in the central gray area of the spinal cord to drive:

1. Motor neurons in the gray area, whose axons exit the ventral root to innervate muscles.

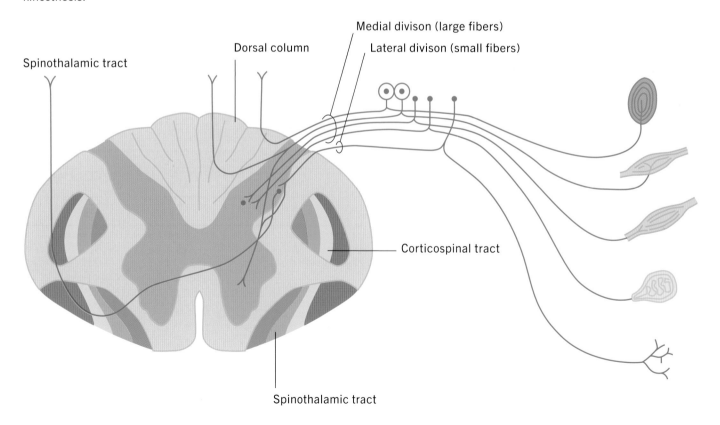

Above: *Projections of sensory dorsal root ganglion cells.*

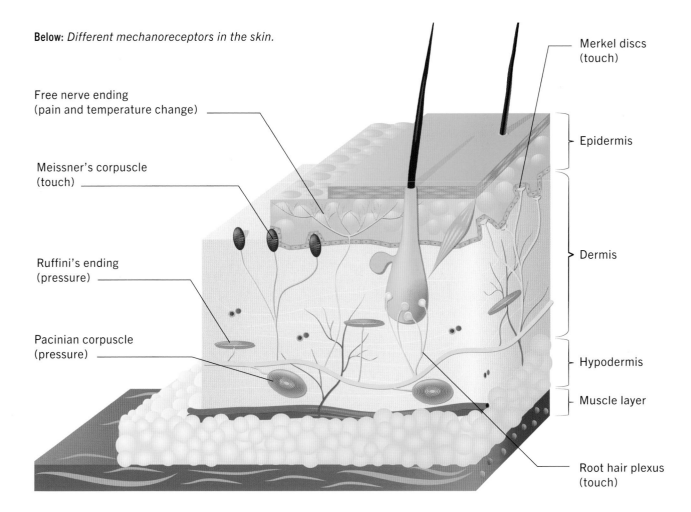

Below: *Different mechanoreceptors in the skin.*

Free nerve ending
(pain and temperature change)

Meissner's corpuscle
(touch)

Ruffini's ending
(pressure)

Pacinian corpuscle
(pressure)

Merkel discs
(touch)

Epidermis

Dermis

Hypodermis

Muscle layer

Root hair plexus
(touch)

2. Interneurons involved in spinal cord control circuits.
3. Transmission neurons that project upward to the brain.

Skin receptors

Touch receptors that detect various kinds of skin indentation or stretch are called mechanoreceptors. There are four major types of mechanoreceptors:

1. Pacinian corpuscles
2. Ruffini endings
3. Merkel's disks
4. Meisner's corpuscles

A fifth type of receptor ending is the so-called free nerve ending. Free nerve endings typically are transducers for temperature and pain.

The four mechanoreceptor types differ in three major ways:

1. Sustained versus transient (temporary) responses.
2. Receptive field size.

3. Frequency response and perceptual result of stimulation.

Receptors that give sustained responses are called "SA" for "Slowly Adapting," while those that respond transiently are called "RA" for "Rapidly Adapting." The different receptor types also have different stimulation frequency ranges and perceptual results.

Small receptive field types have the suffix "1" (e.g., SA1 or RA1), while large receptive field types have the suffix "2." Receptive field size for most skin receptors is a function of its depth in the dermis. The SA1 and RA1 receptors (Merkel disks and Meissner corpuscles) are found closer to the epidermis than the deeper SA2 and RA2 receptors (Ruffini endings and Pacinian corpuscles).

Small receptive field receptors

The activation of Merkel disks codes for the perception of pressure on the skin. These respond primarily to static pressure and to light touches—they have a limited frequency response range of 0.3 to 3 Hz. The other small area receptor is the Meissner corpuscle, which responds transiently (RA1) to frequencies from 3–40 Hz. Activation of this receptor type results in the perception of what is called flutter, such as from a buzzing insect on the skin.

Large receptive field receptors

Ruffini endings are large receptive field receptors that give sustained responses (SA2). They are primarily activated by deformation of the skin when stretched. Ruffini endings respond to a large frequency range from 15–400 Hz. The transient large receptive field receptor type is the Pacinian corpuscle (RA2), which responds to frequencies from 10–500 Hz, the largest such frequency range of any skin mechanoreceptor. Its activation is perceived as vibration.

Given that there is considerable overlap in the frequency ranges of these four mechanoreceptors, the total perception of what is deforming the skin results from a combination of their relative activity, much like the relative co-activation of the three cone types in the retina produces a perceptual range of hundreds of colors.

Pacinian corpuscles

Pacinian corpuscles act as receptors for pressure and vibration. The intrinsic response property of the mechano-receptive ion channel is sustained (SA). However, the onion-like myelin wrapping physically adjusts to constant pressure such that it only transmits deformation to the spike initiation site when the pressure is first applied (it then "adjusts"), and then when the pressure is released. This means that the Pacinian corpuscle only spikes when the pressure is first applied, and when it is released, making its output the transient (RA) type.

The axonal endings of mechanoreceptors contain ion

Below: *The structure of the Pacinian corpuscle. The axon is in red, and the Schwann cell myelin wrappings are in green. The gray, onion-like structure is a specialized type of myelinated wrapping at the end of the axon. The mechano-receptive ion channels are at location B, where the action potential is initiated.*

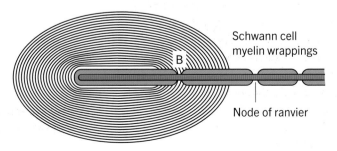

Schwann cell myelin wrappings

Node of ranvier

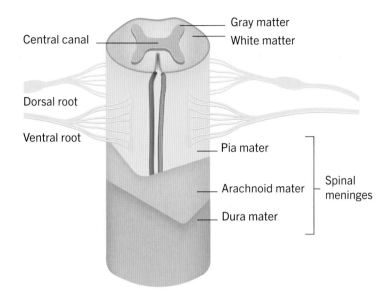

Above: *A spinal cord segment and a dorsal root ganglion.*

channels that are opened by skin deformation. When opened, sodium traverses these channels and depolarizes the axon ending, producing action potentials there that travel toward the dorsal root ganglion and spinal cord.

Total dorsal root ganglion cell input

The axons of the dorsal root ganglion cells in each ganglion project to a particular skin area associated with that spinal cord segment. Afferent axons of the dorsal root ganglia send sensory information to the spinal cord gray area via their entrance of the dorsal root. Motor neurons in the ventral gray area project outward (efferent) to muscles underneath the skin.

Each dorsal root ganglion, and therefore the spinal segment to which it projects, receives inputs from a particular area of the skin called a *dermatome*. When a specific spinal cord segment is severely injured, there is loss of skin sensation in that dermatome, as well as paralysis of muscles beneath that skin area innervated by lower motor neurons from the ventral root of that spinal cord segment.

The density of cutaneous receptors (the number of receptors per skin area) is much greater in the skin of some parts of the body, such as the fingertips, than in other parts, such as the trunk. The area of the skin over which deformation activates a mechanoreceptor is called its *receptive field*. The output of the receptor will be the same regardless of where within the receptive field the stimulation occurs. Even in the case where the receptor is more sensitive in the middle of its receptive field than in the periphery, the output will be nearly the same for a weak stimulus in the center versus a stronger one in the periphery. Thus, the ability to distinguish two points of skin indentation from one depends inversely on receptive field size.

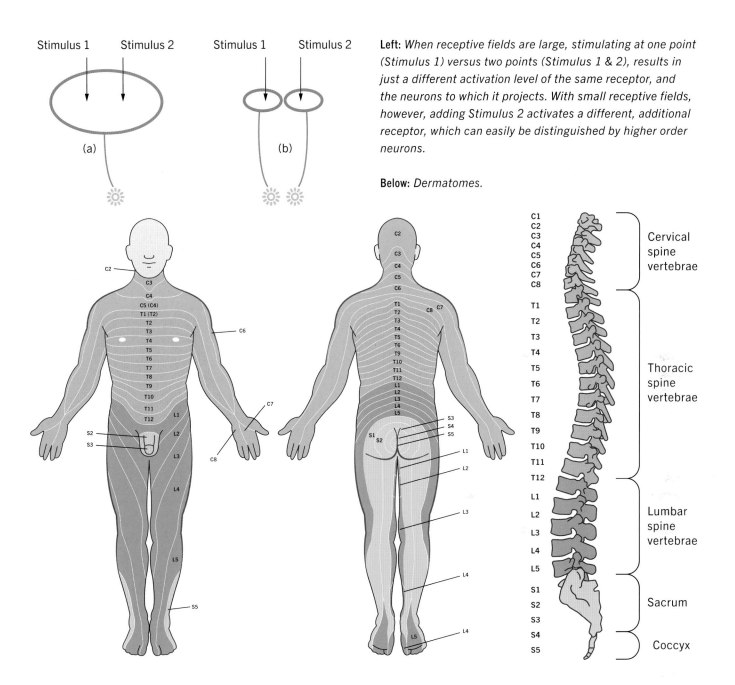

Stimulus 1   Stimulus 2   Stimulus 1   Stimulus 2

(a)   (b)

Left: *When receptive fields are large, stimulating at one point (Stimulus 1) versus two points (Stimulus 1 & 2), results in just a different activation level of the same receptor, and the neurons to which it projects. With small receptive fields, however, adding Stimulus 2 activates a different, additional receptor, which can easily be distinguished by higher order neurons.*

Below: *Dermatomes.*

Cervical spine vertebrae

Thoracic spine vertebrae

Lumbar spine vertebrae

Sacrum

Coccyx

The minimum distance for which one can distinguish one from two stimulus points is about 2 mm on the fingertips and parts of the face, but over 40 mm on the trunk. This is a direct result of the fact that there is a high density of small receptive field cutaneous receptors on the fingertips and face, but a much lower density of large receptive field receptors on the trunk and legs.

Temperature receptors

There are distinct temperature receptors in the skin for heat and cold. Both receptor types are based on small axons that elaborate free nerve endings. So-called warm receptors respond to a range of temperatures above body temperature (37°C), while cold receptors increase firing to colder temperatures. Individual fibers have different peak sensitivities within this range. By comparing the ratios of activity of different fibers with different peak sensitivities, fine discrimination of temperature can be perceived like the hundreds of color perceptions arising from activity in three types of cones. Temperatures higher than those to which the warm receptors respond, and lower than those that activate cold receptors, are transduced by other fibers whose activity is perceived as pain.

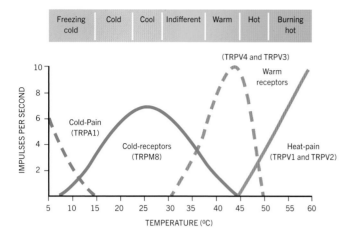

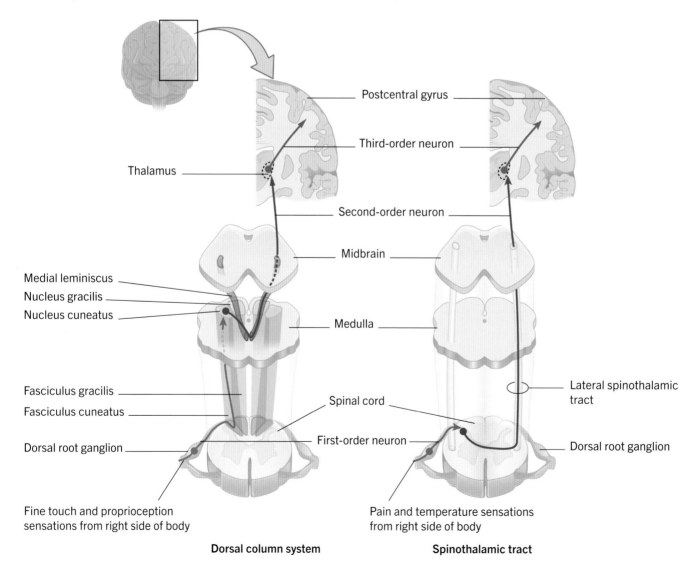

Pain receptors

The perception of pain is induced by a range of stimuli, including extreme temperatures, pressures, and chemical acids and bases that produce tissue damage. The technical name for pain receptors is nociceptors. These receptors are based on small axons with free nerve endings, similar to that of temperature receptors. Pain is different from other cutaneous senses in that its perception is highly context-dependent, and it can result in long-lasting changes in affect, particularly in the case of chronic pain. Pain pathways to and within the brain use unique neurotransmitters, such as endorphins.

Central projections of cutaneous senses

Skin sense information is sent to the brain from the spinal cord gray area by relay neurons along two tracts:

Above: *Population responses of cold, warm, and pain fibers to different temperatures.*

Below: *Projections of skin senses to the brain.*

Dorsal column system

Spinothalamic tract

1. The lemniscal tract, which mediates mechano-sensation.
2. The spino-thalamic tract, which mediates temperature and pain perception.

Cutaneous information reaches brainstem structures such as the medulla as well as the ventral posterior nucleus of the thalamus. The thalamus projects to the somatosensory cortex at the most anterior part of the parietal lobe.

Somatosensory cortex

The map of skin on the somatosensory cortex is known as a homunculus. The homunculus is distorted in two ways:

1. The representation of the skin on the cortex is not proportional to skin area, but to the number of receptors in any skin area. Thus, the representation of the fingertips is nearly the same size as the entire trunk because the number of receptors is about the same in those areas.
2. The head has two representations, one of which fits with its position with respect to the rest of the body, and a second, lateral, enlarged representation.

The cortical representations for pain and temperature from the spino-thalamic pathway are more complex, and not necessarily in register with other skin senses. The spino-thalamic pathway projects directly to the reticular formation so that pain generates autonomic responses such as sweating and tachycardia. Activation of pain receptors also results in activation of the anterior cingulate cortex and the amygdala.

Below: *The somatosensory cortex and homunculus.*

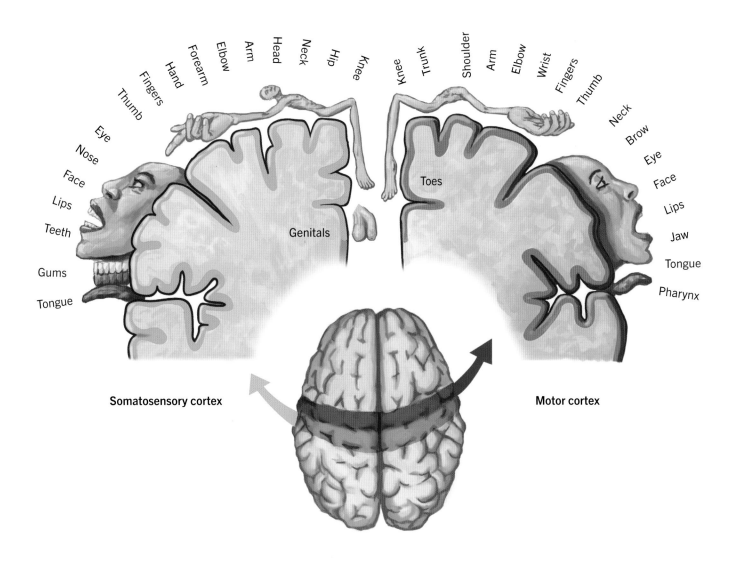

Somatosensory cortex

Motor cortex

// Smell

Smell and taste together, are referred to as chemical senses. Whereas the senses of vision, audition, and somatosensation register the input of energy for perception, the chemical senses of smell and taste detect

Below: Smell pathway in the brain and the composition of olfactory receptors.

molecules that contact specialized receptors in the body.

On each side of the septum in the middle of the nose are three cavities called turbinates. At the top of the superior turbinate is the olfactory mucosa where receptors for odor molecules are located. Above the olfactory mucosa is the olfactory bulb that receives projections from receptor neurons in the mucosa.

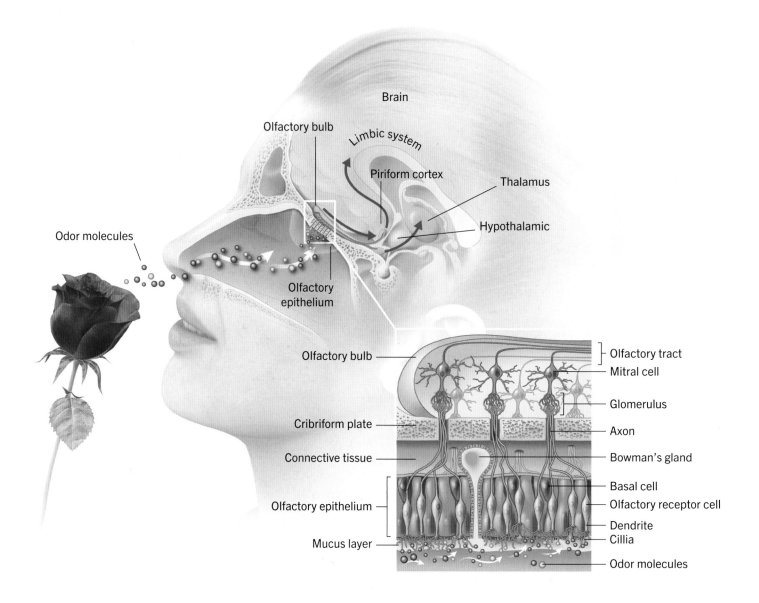

Brain

Olfactory bulb

Limbic system

Piriform cortex

Thalamus

Hypothalamic

Odor molecules

Olfactory epithelium

Olfactory bulb

Cribriform plate

Connective tissue

Olfactory epithelium

Mucus layer

Olfactory tract

Mitral cell

Glomerulus

Axon

Bowman's gland

Basal cell

Olfactory receptor cell

Dendrite

Cillia

Odor molecules

The olfactory bulb and ORNs

The olfactory receptor neurons (ORNs) contain cilia (microscopic hairs) in the mucosa. Each olfactory receptor cell extends 5–10 ciliary processes into the mucosa to detect odorants that reach there. The olfactory receptors are on the cilia in the mucosa. There are about 1,000 different receptor types on olfactory neurons, but a single olfactory receptor neuron expresses only one type of receptor.

The upper side of the ORNs extend axons that project to the olfactory bulb. There are about 5–10 million olfactory receptor neurons, and about 5,000–10,000 receptor cells of each of the one thousand types. Each ORN fires action potentials in response to binding odorants.

The action potentials travel along the ORN axons to the olfactory bulb. This axon bundle is cranial nerve I.

The axon terminals of ORNs release the neurotransmitter glutamate onto the dendrites of cells called mitral and tufted cells in the olfactory bulb. These receptor cells in the olfactory bulb are clustered into about 1,000–2,000 distinct domains called glomeruli. Each glomerulus receives inputs from one receptor type.

Olfactory coding appears extraordinarily complex when compared to color vision. In vision, hundreds of distinct hues can be perceived based on the ratios of activities of three cone types (see pages 76 and 85–6). In olfaction, however, there are 1,000 receptor types, most of which respond to a large subset of all odorants. Because each olfactory glomerulus receives predominant input from about 5,000–10,000 ORNs of a single receptor type, each odorant will produce a unique, complex pattern of activity across the glomeruli in the bulb.

Below: *Olfactory glomeruli and projections via the olfactory tract.*

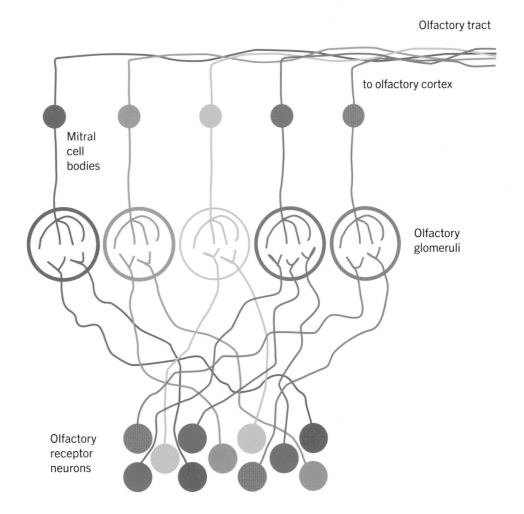

Olfactory tract

to olfactory cortex

Mitral cell bodies

Olfactory glomeruli

Olfactory receptor neurons

There is another olfactory organ at the front of the septum within the nose in most mammals called the vomeronasal organ. Olfactory receptors there mediate the detection of species-specific chemical signaling via what are called pheromones. Pheromones mediate sexual behavioral triggering, such as menstrual synchrony in women living together.

The projections from the olfactory bulb via the olfactory tract reach more areas of the brain directly than other sensory systems. Targets of the olfactory bulb include:

1. The anterior olfactory nucleus.
2. The olfactory tubercle.
3. Pyriform cortex.
4. The amygdala.
5. The entorhinal cortex near the hippocampus.

Of particular note is that there is a direct projection from the olfactory bulb to the pyriform cortex in the temporal lobe that does not involve a thalamic relay, like all other sensory systems.

The direct projection to the amygdala is the basis for the strong emotional affective nature of many smells that can trigger either strong aversion or attraction. The projection to the hippocampal region is related to the rapid learning through the olfactory system. In so-called "single trial learning," a single exposure to a smell (or taste, or combination of both) prior to illness can produce a lifelong aversion to that odor or taste.

In addition to the rapid reaction and learning systems involving the amygdala and hippocampus, there is a secondary system whereby smell reception is combined with taste in the orbitofrontal cortex. This system involves projections from the amygdala and pyriform cortex to the mediodorsal thalamus, which in turn projects to the

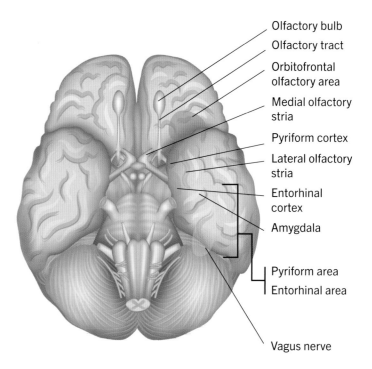

Above: A ventral view of the olfactory areas of the brain.

Below: Olfactory projections to brain areas.

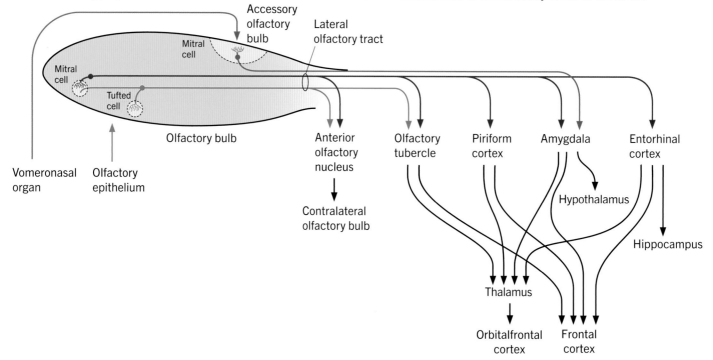

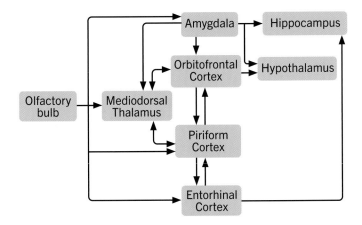

Above: *A ventral view of the olfactory areas of the brain.*

orbitofrontal cortex. Like other sensory system projections through the thalamus, this is the system that generates conscious awareness of smell, and its combination with taste.

The orbitofrontal cortex

Our sense of taste includes a strong contribution from the sense of smell. During chewing, droplets of chewed food travel upwards from the nasal pharynx to the olfactory area. Food particles in these droplets activate olfactory receptors simultaneously with activating taste receptors in the tongue (see pages 104–5).

The smell and taste systems both project to some of the same neurons in the orbitofrontal cortex that give rise to the perception of what we call taste. The sensory combination of smell and taste is referred to as "flavor."

Flavor areas in the orbitofrontal cortex also receive inputs from receptors in the stomach that help regulate hunger. The decrease in hunger as food is consumed by this mechanism is called *alliesthesia*. There is another, faster-acting mechanism of hunger reduction called *sensory-specific satiety*, based on activation of taste receptors in the tongue.

Loss of smell can occur from damage to the olfactory mechanisms in the nose, as well as from several neurological conditions, particularly those that affect the frontal lobe. Loss of smell can occur in Parkinson's disease, Huntington's disease, schizophrenia, and Alzheimer's disease. Down syndrome is also associated with impairment in the sense of smell.

Below: *The orbitofrontal cortex.*

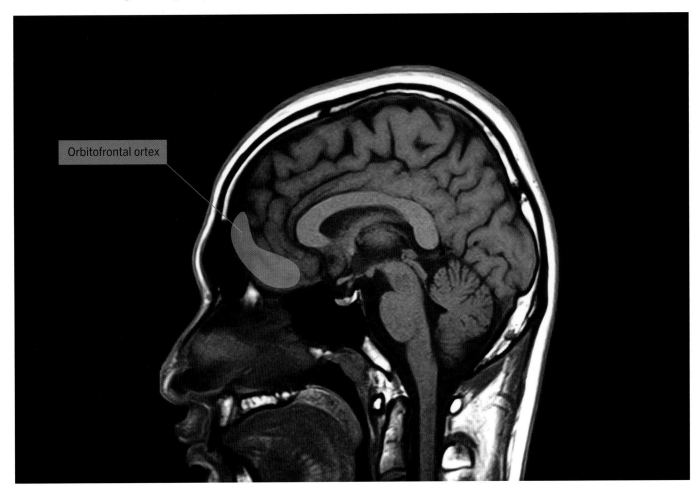

Orbitofrontal ortex

// Taste

Taste has classically been based on the four basic elements of sweet, salt, sour, and bitter. Many researchers now believe that there is a fifth taste called umami elicited by monosodium glutamate (MSG). Most taste receptors are on the tongue. There are also a few taste receptors in the pharynx.

Taste receptors on the tongue are located in structures called papillae that look like little bumps. Most of the papillae on the tongue are called filiform, which do not have taste bud receptors. The three other types of papillae do, however, contain taste bud receptors:

1. Fungiform.
2. Foliate.
3. Circumvallate.

The taste bud contains up to 50 taste receptor cells whose cilia that extend toward the taste pore have chemo-receptors.

Each taste cell within a bud is specific to a single taste element (sweet, salt, sour, bitter, umami), but within each bud there are receptors for different tastes. Receptors for all tastes occur in all three taste bud types. However, there is a greater sensitivity overall at the front of the tongue to the sweet taste, and toward the back for the bitter taste.

The receptor for salt is an ion channel for sodium that allows external sodium to enter the receptor cell and depolarize it. Similarly, the sour taste, which is produced by acidity, is mediated by a channel that fluxes positive hydrogen (H+) ions into the receptor.

The channels for sweet, bitter, and umami are more complex. The binding of the ligand on the receptor protein complex on the outside of the membrane brings about a second messenger cascade inside the receptor cell that mediates its electrical response.

Taste receptors

Taste receptor cells in the taste buds are contacted by the axons of sensory neurons whose cell bodies arise in several

Below: *The structure of a taste bud.*

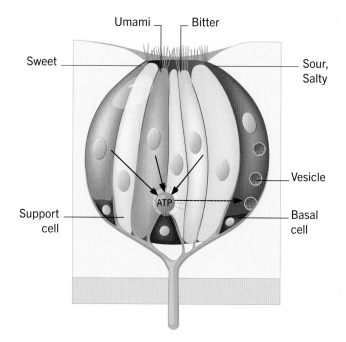

Below: *The gustatory pathway.*

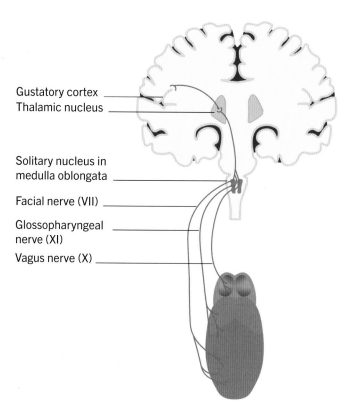

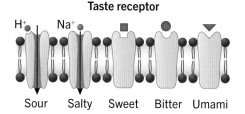

ganglia outside the brainstem. These ganglia include the geniculate ganglion, petrosal ganglion and nodose ganglion. Activation of the taste sensory cells is mediated by the release of glutamate onto receptors on the axons of these ganglia. From there, action potentials travel to several targets in the midbrain, such as the nucleus of the tractus solitaries (solitary nucleus).

Receptors from the front and middle of the tongue project to the brainstem via the chorda tympani nerve. Receptors at the back of the tongue project via the glossopharyngeal nerve, which is cranial nerve IX. A few taste receptors in the pharynx project to the midbrain via the vagus nerve.

The solitary nucleus of the medulla (NST) gathers taste signals and also receives sensory-specific satiety inputs that modulate the activity of projection cells there. Continued consumption of something sweet, for example, reduces the desire to keep consuming glucose. The NST projects to the ventral posterior nucleus (VPN) of the thalamus. The VPN then projects to several brain areas, including the anterior insula-frontal operculum cortex.

Taste cortex

Most taste receptors are activated by multiple substances. A major function of taste cortex is to discriminate between specific tastes based on the ratio of activity of different receptor types. These cortical areas then project to the amygdala and orbitofrontal cortex. The amygdala is involved in learning aversive tastes, or tastes associated with subsequent sickness. Taste projections to the orbitofrontal cortex are combined with olfactory input to produce flavor. There are also projections to the orbitofrontal cortex from the amygdala and the visual system that mediate complex memories about ingested food based on multiple stimulus dimensions relevant to hunger.

Below: *Central projections of taste.*

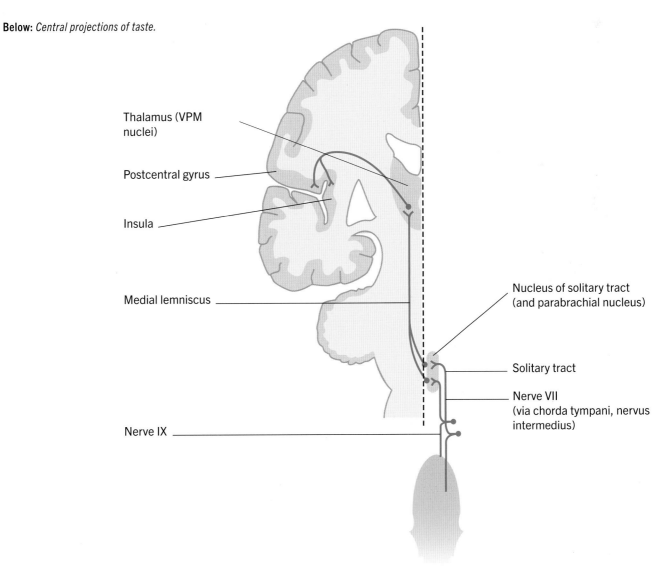

Thalamus (VPM nuclei)

Postcentral gyrus

Insula

Medial lemniscus

Nucleus of solitary tract (and parabrachial nucleus)

Solitary tract

Nerve VII (via chorda tympani, nervus intermedius)

Nerve IX

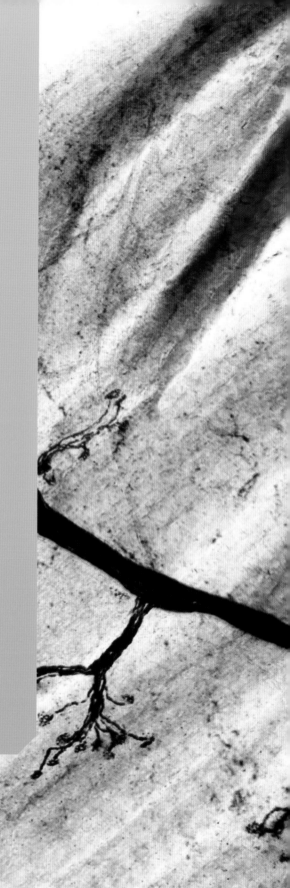

MOTOR SYSTEMS AND ACTIONS

Central nervous systems with brains exist in animals, but not plants, in order to control movement. The choices about where to move are based on sensed information about the environment and on the animal's internal state, such as the existence of hunger. Movement is enabled by muscles whose contraction is mediated by motor neurons. Motor neurons, in turn, are controlled by a hierarchy of mechanisms, ranging from the monosynaptic reflex loop in the spinal cord at the bottom to the primary motor cortex at the top. Structures such as the supplementary motor area, the premotor cortex, the basal ganglia, and the cerebellum all play crucial roles in making movement possible.

A micrograph showing axons of motor neurons (in black) connecting with muscle tissue.

Muscles and motor neurons

As we saw earlier, brains evolved primarily to generate and control movement. Movement is accomplished by muscles that are stimulated by special neurons called motor neurons. Action potentials in motor neurons release the neurotransmitter acetylcholine (see page 65), which binds to receptors on muscle fibers, opening depolarizing ion channels, and causing a muscle action potential. The elevation of calcium concentration during the muscle action potential makes the muscle fibers contract. Thus, motor neurons translate nerve signals into mechanical actions—the change in length and tension of muscles.

The cell bodies of motor neurons that control muscles in the limbs are in the ventral spinal cord gray area. Their axons exit the ventral root, pass through the dorsal root ganglion (which contains sensory neurons), and proceed to the muscle.

At the muscle, the axon branches several times and provides synaptic endings at the end of each axon terminal. Generally, each axon terminal innervates one muscle cell, and is the sole driver of that muscle cell. A single motor neuron drives as many muscles cells (usually on the same muscle) as it has axon terminals. This collection of terminals is called a *motor unit*.

Below: *The somatic nervous system.*

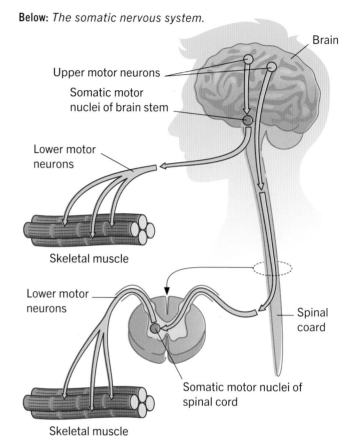

Brain

Upper motor neurons

Somatic motor nuclei of brain stem

Lower motor neurons

Skeletal muscle

Lower motor neurons

Spinal coard

Somatic motor nuclei of spinal cord

Skeletal muscle

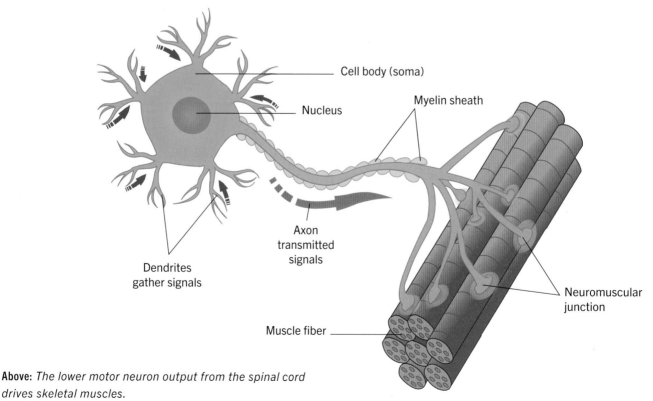

Cell body (soma)

Nucleus

Myelin sheath

Dendrites gather signals

Axon transmitted signals

Neuromuscular junction

Muscle fiber

Above: *The lower motor neuron output from the spinal cord drives skeletal muscles.*

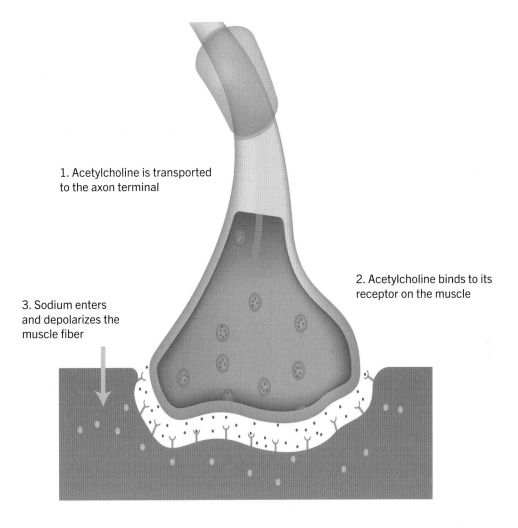

Right: *A detailed view of a single axon terminal and the underlying receptor area of the muscle cell.*

1. Acetylcholine is transported to the axon terminal

2. Acetylcholine binds to its receptor on the muscle

3. Sodium enters and depolarizes the muscle fiber

The motor neuron synapse

The motor neuron axon terminals synapse on muscle cells at locations called *junctional folds*. Receptors for acetylcholine located there flux sodium and a lesser amount of potassium, causing the membrane near the receptor to depolarize (become less negative with respect to the outside). The depolarization event is called the end plate potential (EPP). Deep within the junctional folds are voltage-gated sodium channels. When the junctional fold membrane is depolarized, these voltage-gated sodium channels further depolarize the membrane in a positive feedback loop that brings about the muscle cell action potential.

Generally, each spike at a motor neuron axon terminal elicits a single muscle cell axon potential, which causes a small contraction of the muscle via a calcium mechanism. The force of the muscle contraction depends on the rate of action potentials in each muscle cell as well as the total number of muscle cells activated by different motor neurons.

The muscle cell is composed of different kinds of fibers—thick and thin filaments. The thick filaments are made of myosin, while the thin filaments are a complex of F-actin, troponin and tropomyosin. Calcium released from intracellular stores by the muscle action potential acts on troponin to produce muscle contraction.

At rest, the troponin binding sites are blocked. However, when calcium ($Ca++$) binds to troponin, it causes tropomyosin to change its form and expose the myosin binding sites on the actin. The thick filament myosin head then binds to the actin site, which now rotates, sliding the thin filament to the left. This rotation causes the adenosine diphosphate (ADP) molecule on the myosin head to detach. The muscle contracts via the sliding of the thick versus thin filaments. The myosin head then binds an ATP molecule where the ADP molecule once was bound that resets the myosin head to its original position. Note that the resetting process after the sliding contraction mechanism requires energy via ATP. In rigor mortis and several other conditions, for example, where energy stores in the body are depleted or unavailable to muscle cells, constant contraction is sustained.

As long as calcium and energy are still present in the muscle fiber, the binding and unbinding process repeats

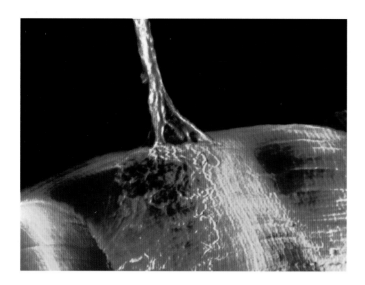

Above: *Motor neuron synapse, SEM photo.*

itself, and the myosin filaments move step by step along the length of the actin filaments and the muscle fiber continues to contract. When the motor action potentials cease, the muscle fibers relax because calcium is taken up by the internal organelles. Without calcium, troponin blocks the interaction of myosin and actin, and the thin and thick filaments slide apart, and the muscle fibers return to their resting length.

How muscles work

Muscles only generate force when they contract. Thus, to move a limb about a joint such as the elbow in extension and flexion, muscles are arranged in antagonistic pairs. The motor neuron spinal cord output is driven by multiple pathways. Some of these pathways are reflexive, driven by spinal cord and brainstem mechanisms, while others, mediated by motor cortex in the brain, control voluntary movement.

The amount of force generated by the muscle (strength and speed of muscle contraction), is controlled by:

1. Motor neuron firing rate.
2. The type of muscle fibers that are activated.

There are three main types of voluntary muscles:
1. Type I slow twitch (ST).
2. Type IIa fast twitch (FT) fatigue resistant (FR).
3. Type III (FT) fast fatiguing (FF).

They vary by size, order of recruitment, amount of force generated, and fatigability depending on the need of the body for different amounts of muscle force to be exerted for different amounts of time.

Type I muscle fibers are in muscles that control normal posture, such as standing up. People can stand up for hours, and, in doing so, do not need to generate high muscle force (compared to running, for example). They are thus called slow twitch. These muscles use standard oxidative metabolism, that is, they use oxygen delivered by the blood stream to obtain energy from sugars and carbohydrates, usually via the intermediate energy-storing molecule, ATP (adenosine triphosphate).

Type IIa muscle fibers are called fast twitch (FT) because they generate higher forces than Type I slow twitch fibers. The red meat sold in grocery stores is mostly this type of muscle. They are called fatigue-resistant (FR) because they can exert force for a moderate amount of time, less than Type I fibers, but more than Type III fibers. The metabolism of Type IIa fibers is primarily oxidative, however, they can continue into oxidative debt using the glycolysis mechanism. Glycolysis is a metabolic process that does not require oxygen to convert glucose into pyruvic acid to create ATP (adenosine triphosphate) for muscle operation.

Type III muscle fibers generate the highest force, for the shortest period of time. These typically have a white color. They are quickly fatigued and have the highest use of the glycolysis mechanism. When one engages in high force, long-term activity, such as distance running, the leg cramps that can be experienced arise from the buildup of glycogen in the muscles from this metabolism.

Generally, the highest force muscle fibers such as Type III are driven by the largest diameter motor neurons. These in turn tend to be driven by upper motor neurons from the motor cortex in the brain. The slow twitch muscles are driven by smaller diameter motor neurons, typically under control of homeostatic posture maintenance circuits in the spinal cord and brainstem.

Below: *Recruitment of different muscle fiber types.*

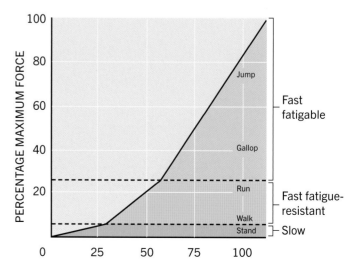

Fast fatigable fibers can only contract for a short period of time (about a minute), whereas fast fatigue-resistant fibers can fire for several minutes, and slow twitch fibers can fire indefinitely.

Although most muscles are predominately composed of a single fiber type, many have multiple fiber types that allow the muscle to generate either high force for a short period of time, or lower force for longer periods.

The different muscle types are driven by different motor neuron types from the spinal cord that are activated by different control circuits there.

Stretch receptors

The central nervous system receives several types of input about the status of muscle contraction and limb position. Two of the most important are:

1. Golgi tendon organs.
2. Muscle spindle stretch receptors.

Golgi tendon organs are a type of proprioceptor located at the junction of muscles and tendons (which connect muscles to bones) that measure muscle tension. Golgi tendon organs prevent over-contraction of muscle, and allow fine control and the limiting of tension when grasping delicate objects, such as when one grasps and picks up an egg.

The other major type of proprioceptors is the muscle spindle. The receptor for the muscle spindle is also the axon ending from a neuron whose cell body is located in the dorsal root ganglion. The axonal ending of the spindle responds to muscle stretch when the muscle is contracted by firing off the corresponding alpha motor neuron from the ventral root of the same spinal cord segment.

Because the muscle contraction state controls limb position, the muscle spindle reports limb position via its state of contraction and is thus a proprioceptor. Similarly the rate of contraction (its first derivative) indicates the rate of limb movement, a sensory construct referred to as kinesthesis.

Below: *The stretch reflex and the golgi tendon.*

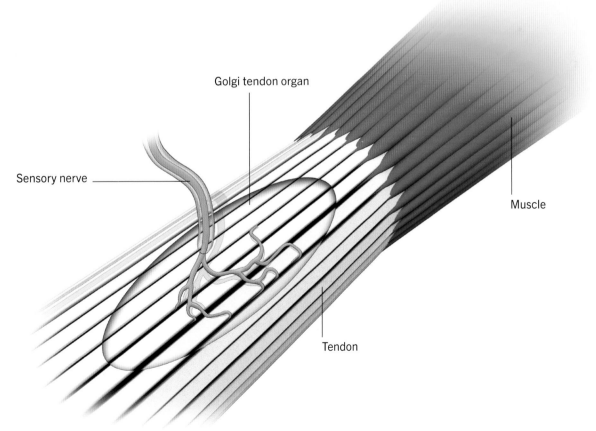

Golgi tendon organ

Sensory nerve

Muscle

Tendon

// Brainstem, spine, and movement

Motor control uses many negative feedback mechanisms. A good illustration of how negative feedback works is heating your house in the winter. Suppose you have a furnace, but no thermostat. As the house gets cold, you turn on the furnace. The house heats up and eventually gets too hot, so you turn off the furnace. This is a single level of negative feedback control. You are both the temperature sensor, and heat controller (via operating the furnace). Your actions in turning the furnace on or off reduce the difference between the temperature you desire for the house and the

actual temperature. Non-mammalian vertebrates regulate temperature in a similar manner, behaviorally, by moving to a warm spot when too cold, or a cold spot when too warm.

But mammals add the equivalent of a thermostat to the system. The thermostat has a device for measuring temperature, and a switch controlled by that device that turns on the furnace below the temperature set on the thermostat and turns off the furnace when the temperature is above that setting. The thermostat now accomplishes the negative feedback control by reducing the difference

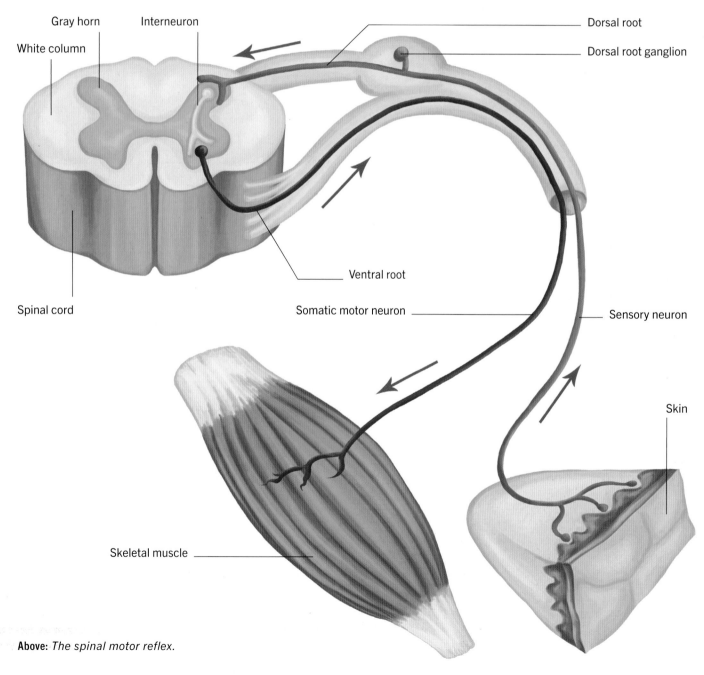

Above: *The spinal motor reflex.*

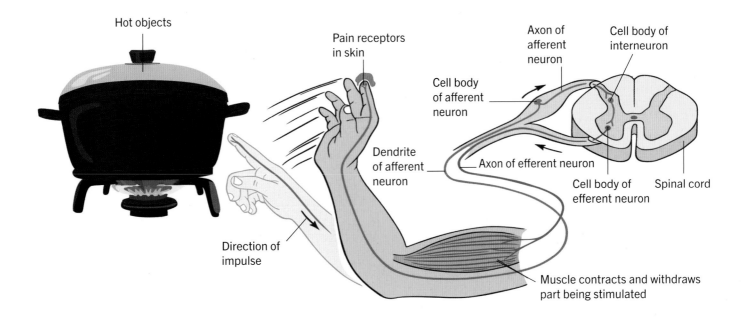

Labels: Hot objects; Pain receptors in skin; Axon of afferent neuron; Cell body of interneuron; Cell body of afferent neuron; Dendrite of afferent neuron; Axon of efferent neuron; Cell body of efferent neuron; Spinal cord; Direction of impulse; Muscle contracts and withdraws part being stimulated

Above: *The withdrawal reflex.*

between the set temperature and the actual temperature in the house.

Because the brain has multiple levels of control, it is a hierarchical control system. What does this mean? The fact that the thermostat controlling the furnace is settable is what makes it a hierarchical control system. The lowest level is the thermostat turning the furnace on when the temperature is below the set point, turning it off when above. The next hierarchical level is the temperature set point on the thermostat. You might set the thermostat to a lower temperature when you are busy in the middle of the house compared to when you are sitting by a window, where it is colder than the middle of the house. Some smart thermostats have sensors to know where people are located within a house, and set the thermostat set point accordingly. This is another hierarchical level.

Spinal reflexes

Spinal reflexes for constant extension are negative feedback control systems mediated by the central nervous system. We can see how this works by looking at a spinal cord control loop involving the quadriceps muscle.

This muscle is driven by a lower, or alpha motor neuron (blue) whose cell body is located in the ventral spinal cord gray area. The spinal cord receives feedback about the state of stretch of this muscle from muscle spindles that induce spikes in Ia afferent fibers that synapse in the spinal cord. When a person is standing, the quadriceps is required to adjust its tension based on fatigue in the muscle, changes

in upper body position and other factors. If the knee starts to buckle, the muscle is stretched, and the firing of the muscle spindle Ia afferent increases. The Ia afferent synapses directly on the lower motor neuron that drives this muscle, causing it to contract harder and return to the set position for standing upright.

All this is done without conscious awareness. The spinal control of muscle tension to maintain constant muscle length is a negative feedback control system. The muscle spindle reports the difference between the desired length and the actual length (error), and synapses on a motor neuron to return the muscle to the correct length, reducing the error.

If someone wants voluntarily to kneel, the drive from upper motor neurons forces the knee to bend, overriding the lower level reflex. There are also inputs to the lower, alpha motor neuron from the brainstem that adjust the quadriceps tension to achieve balance and for gait control.

Another type of reflex driven by spinal cord mechanisms causes a limb to move, rather than maintain a set position. If a person touches a hot pot with their hand, pain receptors in this skin are activated that synapse on the biceps flexion muscle, causing the lower arm to move away from the candle flame.

Consider the case where, instead of the hand touching a flame, a barefoot person steps on a sharp object. In this case the pain withdrawal reflex must override the static limb position reflex to cause withdrawal of the lower leg. Withdrawal of the lower leg throws the person off balance, requiring the other leg and upper body to compensate.

This compensation is based on spinal cord and brainstem mechanisms at a higher level of the motor control hierarchy.

Movement sequences

Spinal reflexes can be "chained" together to create coordinated movement sequences, such as walking. Walking consists of a cyclical chain of events produced by the central pattern generator in the spinal cord, using information such as when each limb hits the ground and the extension state forward or backward of that limb. This movement sequence is known as a "gait."

Starting with your feet together, the sequence of actions for walking has these steps:

1. Lean forward and to your right (a voluntary act).
2. Swing your left foot forward to catch yourself from falling (a reflex).
3. Swing your right arm forward, and transfer your weight to your left leg (a reflex).
4. Swing your right foot and left arm forward (a reflex).
5. Repeat (a reflex).

Most of the control neural circuitry for standard gaits like walking and running exist within the spinal cord itself (distributed across segments for both the arms and legs), and can be carried out without brain involvement. This was shown clearly in a famous experiment of the Nobel prize-winning neuroscientist Charles Scott Sherrington. A cat, paralyzed because of injury to its cervical spinal cord, walked normally, with alternating fore and hind limbs, when placed on a moving treadmill. The movement was accompanied by muscle spindle stretch receptors in each limb, such that, when the limb reached the rearmost position, weight was transferred to the other limb and the rear limb advanced forward.

Spinal cord mechanisms within cervical spinal cord segments deal with weight transfer and limb alternation of the fore limbs, while mechanisms within lumbar segments do the same for hind limbs. Neural circuits extending most of the length of the spinal cord coordinate counter-phase movement between the fore and hind limbs.

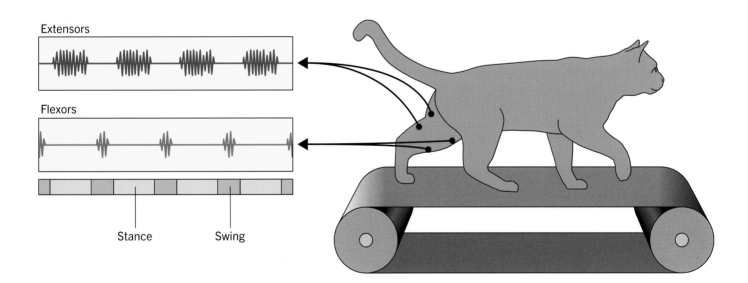

Extensors

Flexors

Stance   Swing

Above: *Sherrington's cat experiment.*

Opposite: *Charles Scott Sherrington.*

The role of the brain

Although spinal cord mechanisms for controlling a standard gait are adequate for walking on a constant speed treadmill, the real world upon which we walk is more complicated. There are holes and hills and obstructions that require changes in gait and movement direction. The brain deals with these complexities by using sensory input from outside the spinal cord, namely from the brain. Of particular importance for locomotion are the vestibular and visual senses.

The brainstem

At the top of the spinal cord is the brainstem (see pages 48–52). The brainstem constitutes a hierarchical level of movement control above that of the spinal cord.

Specifically, the brainstem adds visual and vestibular sensory information from higher brain levels to spinal cord control mechanisms. The brainstem also gives rise to an important motion control structure called the cerebellum (see page 50) that fine-tunes movement sequences.

The red nucleus of the midbrain is involved in gait control. This nucleus receives motor cortex input and is part of the control system for the pace of walking or running. The axon tract from the red nucleus to the spinal cord is called the rubro-spinal tract.

Visually guided balance uses information from the eyes about self-movement to maintain balance. This tract originates in the superior colliculus (see page 83), which receives visual input directly from retinal ganglion cells in the

Below: *Brainstem motor control areas.*

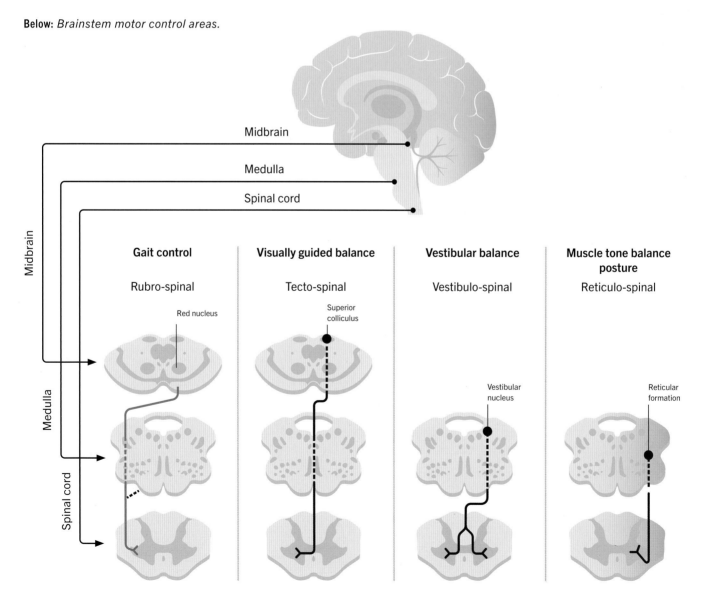

Midbrain

Medulla

Spinal cord

Midbrain

Medulla

Spinal cord

| **Gait control** | **Visually guided balance** | **Vestibular balance** | **Muscle tone balance posture** |
| Rubro-spinal | Tecto-spinal | Vestibulo-spinal | Reticulo-spinal |

Red nucleus

Superior colliculus

Vestibular nucleus

Reticular formation

eyes. The tract is called the tecto-spinal tract. If one steps in a hole while walking, the body will suddenly lean away from the vertical. The visual detection of this movement is transmitted by the tecto-spinal pathway and causes muscles to extend in compensation on the side toward which one is leaning.

The semicircular canals in the inner ear that detect body

Above: *Spinal cord mechanisms may be sufficient for walking on a treadmill, but the real world has many obstacles that require the use of the brain.*

movement and rotation project to the vestibular nucleus in the medulla. This nucleus projects via the vestibule-spinal tract to the spinal cord. This system uses semicircular canal detection of tilt or rotation to accomplish balance, similar to that of the tecto-spinal system.

Another tract from the medulla originates in the medullary reticular formation. The reticular formation receives input from a variety of senses and projects to the spinal cord via the reticulo-spinal tract.

The cortico-spinal tract originates in the primary motor cortex, and is the main means for controlling voluntary and learned movements, particularly those involving the upper extremities such as hands and fingers. People are not born with the ability to tie a knot, skip or dance the tango. These movement sequences are learned and controlled by mechanisms in the frontal lobe.

The cerebellum

The pons is crucial for motor control because it is the input/output area for the cerebellum. The cerebellum is necessary for:

1. Movement coordination across several joints.
2. Fine control of movement for any one joint.
3. Learning and adjusting movement sequences to improve performance.

The cerebellum is an amazing brain structure in several ways. One of these is that the number of neurons in the cerebellum is in the order of, and perhaps larger than, the number of neurons in the entire rest of the brain. The fiber tracts to and from the cerebellum (called peduncles) are massive. The cerebellum has three main divisions: the vestibulocerebellum, the cerebrocerebellum and the spinocerebellum.

The vestibulocerebellum

The smallest part of the cerebellum is the vestibulocerebellum. This area receives inputs from the vestibular system, and projects to brainstem and spinal cord neurons to add the detection of balance to spinal cord

Below: *Regions of the cerebellum.*

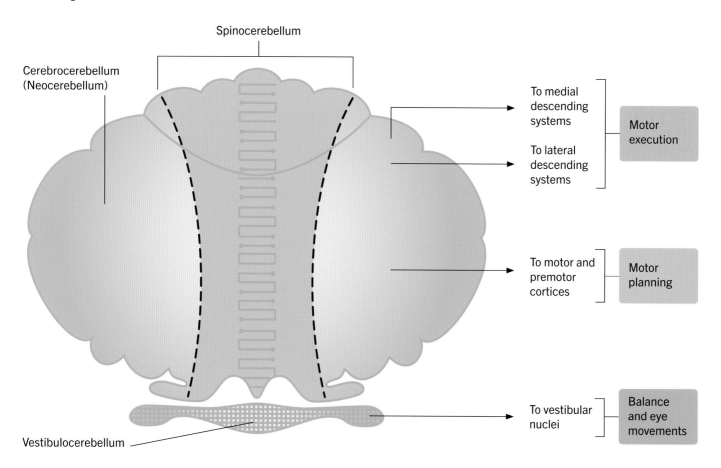

control. If you are bumped and start falling forward, for example, you will transfer weight to the ball of your feet and extend your arms forward. The vestibulocerebellum is also involved in eye movement coordination. One example of this is your ability to fixate on some external object while rotating your head, with the eyes moving in the opposite direction of the head to maintain its fixation.

The cerebrocerebellum

The cerebrocerebellum (neocerebellum) is involved in coordinating voluntary control of multiple limb and muscle groups.

The spinocerebellum

The spinocerebellum in the medial part of the cerebellum that is most responsible for coordination of skeletal muscles. All parts of the cerebellum participate in what is called feed-forward control. For the arm to reach a target, it must rotate both joints at a particular relative rate so that the end of the upper arm hits the target during the double joint rotation. The essential point is that there is no receptor in the body

that can detect when either rotation is incorrect, or which rotation is incorrect if the target is missed. The brain must compute how to move the two joints to achieve arm extension to a given position. This requires an internal "model" of the limbs and joints that allows prediction (feed-forward) of the limb endpoint position from rotation about the joints. This coordinated movement:

1. Must be learned with practice.
2. When learned, must include a "model" of the arm and joints such that, given a visual input about where the target is located, the brain (cerebellum) can compute the appropriate relative rotation rates of the two joints.

Note that, of course, the relative joint rotation rates are different for every target location and rate of reaching. They also change if the shoulder anchor of the upper arm changes. Once the feed-forward model is learned by the cerebellum, you can, for example, touch your nose with your fingertip with your eyes closed, or put another way, with no visual information whatsoever.

Below: *Feedforward control.*

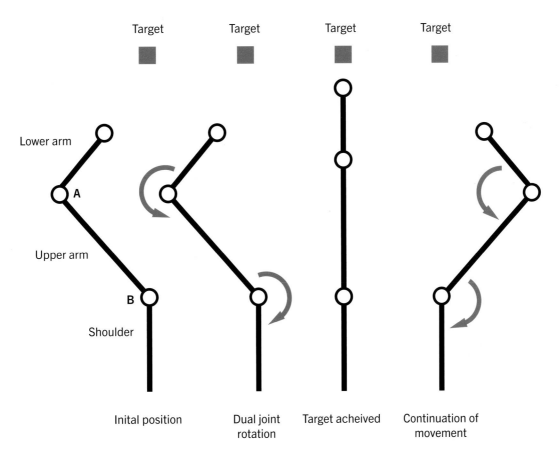

Brain circuits that control movement

Above the brainstem are the subcortical and cortical areas that control movement at higher levels. The neocortex consists of four lobes—parietal, temporal, occipital, and frontal. Three of the four lobes of the neocortex (parietal, temporal and occipital) are primarily sensory. The frontal lobe, which constitutes about half of the entire brain, controls movement and behavior.

The area of the neocortex just in front of the central sulcus is called the primary motor cortex (M1). The primary motor cortex is the final common output of all the computation for movement done by the rest of the frontal lobe, after receiving sensory information from all the rest of the lobes. Neurons in layer V of the primary motor cortex send their axons down to targets in the spinal cord via the cortico-spinal (pyramidal)

Below: *A left side view of the human neocortex highlighting some cortical areas involved in motor control.*

tract. The cortico-spinal tract runs parallel to the other, so-called non-pyramidal tracts that arises in the brainstem.

The cortico-spinal tract represents yet another level in the motor control hierarchy. It is larger in relation to the non-pyramidal tracts in primates compared to other mammals, and larger in humans than in other primates. This tract mediates behaviors that are part of learned, complex, highly planned behaviors that are characterized by complex decision making and contingencies. The large frontal and prefrontal cortical areas are the sites of complex goals, plans, and contingent decisions.

The axis of abstraction

Within the frontal lobe lies what has been termed an axis of abstraction. This axis proceeds from the most abstract representation of a goal in the prefrontal cortex (see page 55), to the motor output that drives specific muscles in

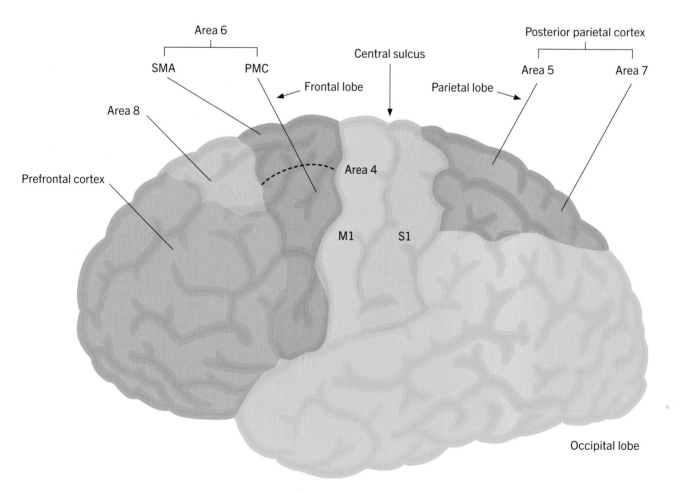

Area of body control

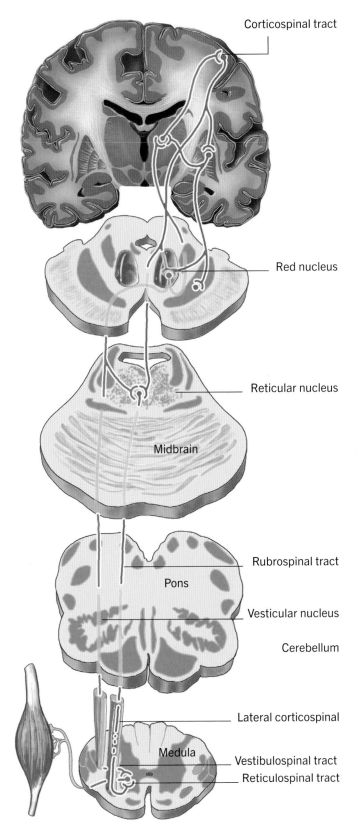

Corticospinal tract

Red nucleus

Reticular nucleus

Midbrain

Rubrospinal tract

Pons

Vesticular nucleus

Cerebellum

Lateral corticospinal

Medula

Vestibulospinal tract

Reticulospinal tract

Spinal cord

Left: *Brain motor cortex pathways.*

primary motor cortex. It allows the conception of plans that both have more steps to completion, and more complex steps. This increased sophistication in planning supports tool making and use, and complex social interactions dependent on differential, individualized relationships between members of a tribe or group.

The highest level of abstraction is the prefrontal cortex, the most anterior region of the frontal lobe. This area of the brain tends to be larger in primates than other mammals, and even larger in humans than in most primates. The pursuit of goals that require many steps and are to be maintained despite distractions requires the representation of the goals and current progress state in the dorsolateral prefrontal cortex, a major component of the working or short-term memory system.

At the other end of the axis of abstraction is the primary motor cortex, just anterior to the central sulcus. Between these two ends of the axis is the representation of "ways and means" of accomplishing goals, from the general to the particular.

Learning and voluntary motor control

Upper motor neurons in the primary motor cortex control single muscles via synapses on lower motor neurons in the ventral root of the spinal cord. Two higher order areas are responsible for controlling associated groups of muscles, such as flexing all the fingers into a fist—the supplementary motor area (SMA) and the premotor cortex (PMC). They are distinguished by their dependences on learning.

The premotor cortex controls muscle groups when new movements are made, or during the learning of movement sequences. Making new movements or learning movements requires:

1. Sensory input about the state of the body or body part (hand for example) with respect to some target or goal.
2. A memory representation of what is desired to be accomplished.
3. Activation of the cerebellum to lay down circuits for precise and coordinated control of movement sequences.

The areas of the brain that tend to be activated with PMC controlled movement include the lateral prefrontal cortex (which provides working memory), the parietal cortex (which provides sensory input relative to movement such as visually guided behavior), and the cerebellum (for coordination).

When a motion sequence has been well learned, such as

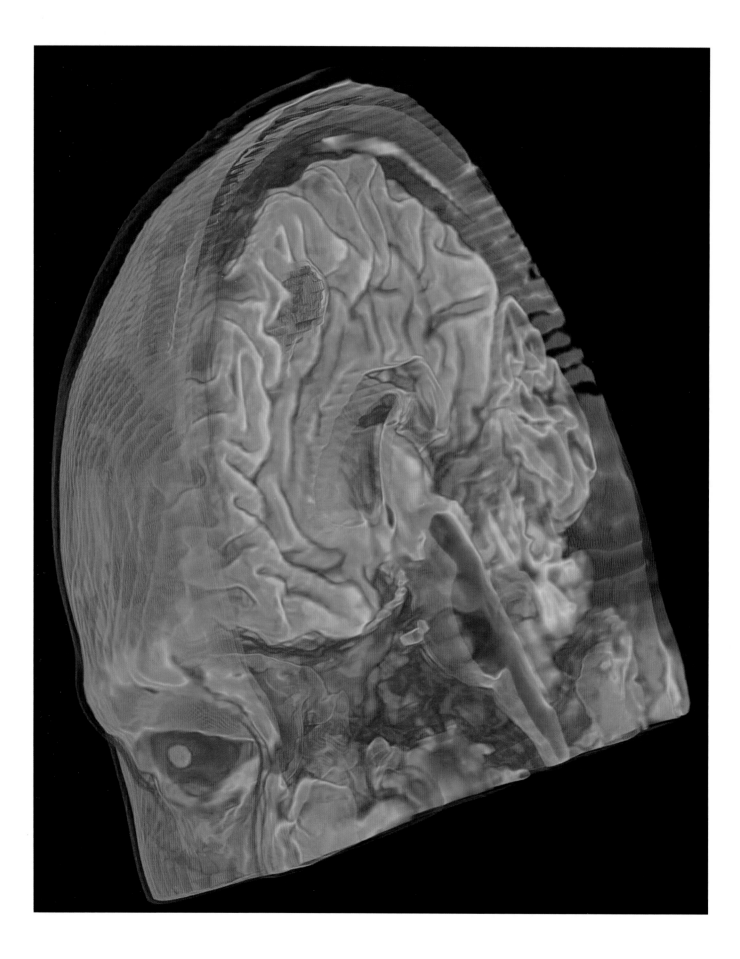

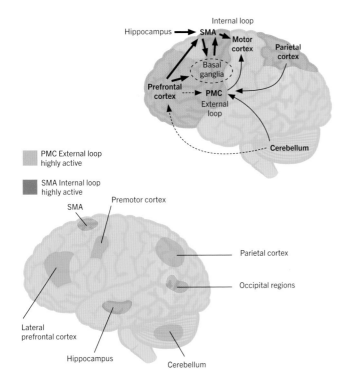

PMC External loop
highly active

SMA Internal loop
highly active

Above: *The loops of the premotor cortex (PMC) and the supplementary motor area (SMA).*

Oppposite: *Brain activation during voluntary action. The red area is the supplementary motor area (SMA).*

Below: *The basal ganglia.*

playing a piano piece, it can be executed without continuous sensory input. In these cases, the supplementary motor area is responsible for controlling the movement. The SMA includes the hippocampus (for access to long-term memory) and some occipital regions for general monitoring of the movement sequence. SMA activation is associated with the initiation of almost all voluntary movements. The SMA will even be activated if a person simply imagines making a movement sequence.

For any given goal, there are a variety of methods available for accomplishing that goal. Each method may also have alternative sub-methods. These various methods and sub-methods are stored in the frontal lobe neocortex. This information is organized in a map in which each method, once activated, automatically activates all the relevant sub-methods.

Proceeding from an abstract goal, such as having dinner, to the specific sequence of actions necessary to get dinner, requires choosing a path through all the methods and sub-methods from the abstract goal all the way to the final motor output in primary motor cortex to accomplish that goal. The choices along this path are made by the basal ganglia.

The basal ganglia

The basal ganglia are subcortical structures located just lateral to the thalamus on each side of the brain below the neocortex (see page 55). The basal ganglia are comprised of two input structures, the caudate and putamen, together

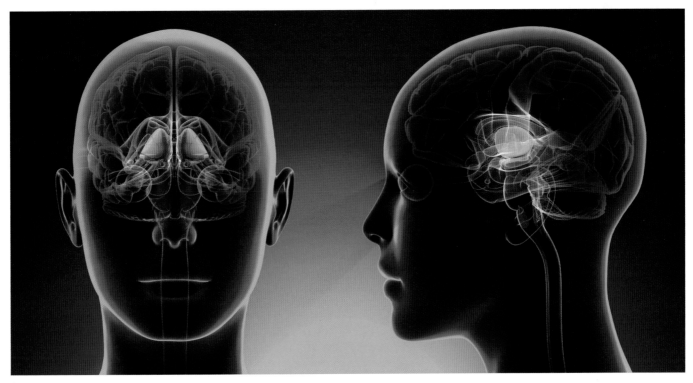

called the striatum. The output structure of the basal ganglia, which projects to the thalamus, is called the globus pallidus.

The basal ganglia receive broad sensory and memory input from the entire neocortex, and project their output to the ventral anterior (VA) and ventral lateral (VL) area of the thalamus. These thalamic areas project, in turn, to the frontal lobe. The basal ganglia are crucial for both the initiation of movement and the modification of movement sequences to deal with obstacles of various kinds.

When you decide to have dinner, the idea goes through several stages in the brain. The high-level goal of having dinner arises in the prefrontal cortex, which activates the alternatives of eating at home, ordering meal delivery, and eating at a restaurant. This goal then activates the basal ganglia, which also receive sensory and memory input about the relevant current state of the world with respect to that

Below: *The basal ganglia work to activate an action pathway through the thalamus to the frontal lobe, going from the abstract goal of having dinner to the specific means to achieve that goal of riding an electric bike to a restaurant.*

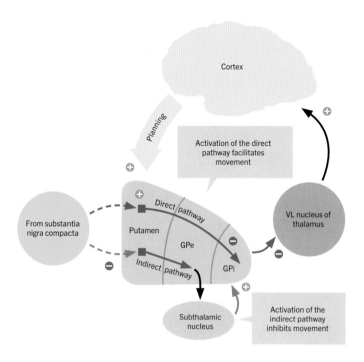

Above: *Direct and indirect circuits of the basal ganglia.*

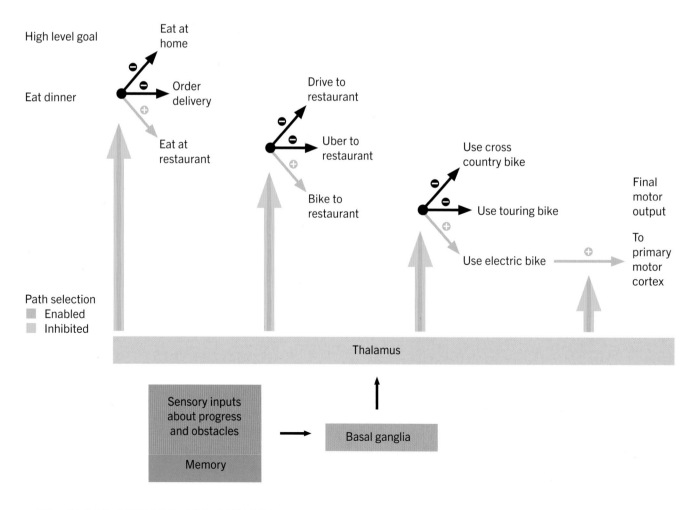

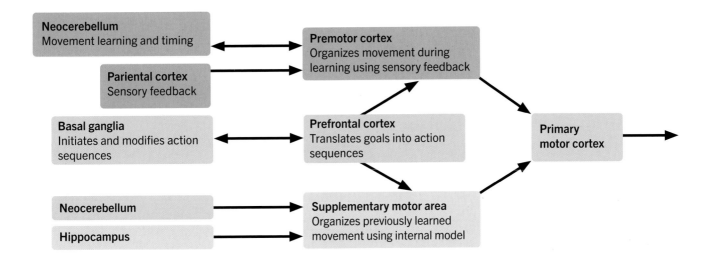

Neocerebellum Movement learning and timing	Premotor cortex Organizes movement during learning using sensory feedback
Pariental cortex Sensory feedback	
Basal ganglia Initiates and modifies action sequences	Prefrontal cortex Translates goals into action sequences	Primary motor cortex
Neocerebellum	Supplementary motor area Organizes previously learned movement using internal model
Hippocampus	

goal. Perhaps you remember that there is no food in the refrigerator, so eating at home won't work. Perhaps the food delivery service has been bad lately, so you don't want that either. The basal ganglia output then selects "eat at restaurant" and inhibits all the other alternatives through its projection to the thalamus.

Once the "eat at restaurant" alternative is selected, it activates various ways to get to the restaurant. Alternatives activated in the frontal lobe might be to drive to the restaurant, to take an Uber, or cycle to it. Once again, the basal ganglia, with sensory and memory input, choose one of these options. This process proceeds all the way to the frontal lobe output, the primary motor cortex to control riding the electric bicycle.

The basal ganglia receive information from:

1. The frontal lobe, about the overall goal and the pathway being taken to the goal.
2. The parietal, temporal, and occipital sensory cortex lobes and subcortical projections, about progress and obstacles relevant to the goal.

The basal ganglia project two outputs to the VL thalamus from the interior segment of the globus pallidus:

1. An excitatory output via what is called the direct pathway that activates one choice among all the alternatives activated in the frontal lobe.
2. Inhibitory outputs via the indirect pathway that inhibit all non-chosen alternatives in the frontal lobe pathway.

Cerebrocerebellum

The cerebrocerebellum (or neocerebellum) coordinates and fine-tunes frontal lobe movement circuits in much the

Above: *A schematic of frontal lobe motor control with PMC generated movement in red, and SMA control in blue.*

same way that the spinocerebellum does for movements generated by the spinal cord and brainstem. The cerebrocerebellum receives input from the PMC pathway during learning to program multi-joint coordination and smoothing. As the motor sequence is learned, the cerebrocerebellum output adds control to the PMC loop. After the control of the motor sequence becomes well learned, the programmed cerebellum participates in the SMA loop.

MIRROR NEURONS

An unusual class of neurons called "mirror neurons" were discovered by the neuroscientist Giacomo Rizzolatti and his colleagues in the last quarter of the 20th century. Although these neurons were initially discovered in the monkey ventral premotor cortex, they have been found in other areas as well, and in many other animals, from birds to humans. The remarkable property of mirror neurons is that they not only fire when an animal performs certain acts involving a representation in that cortical motor area (as expected), but they also fire when the monkey observes someone else performing a similar act. The function of mirror neurons is unclear. Researchers have suggested that they may be involved in learning, imitation, and even empathy. Their discovery is exciting because it suggests that neural circuits in the frontal lobe and elsewhere may have more sophisticated properties than merely being part of a clockwork motor control mechanism.

// Major motor dysfunctions

Motor function can be impaired by dysfunction in muscles, motor neurons, synapses between them, or from higher-level central nervous system disorders.

Disorders of muscles and motor neurons

Muscular dystrophy. Muscular dystrophy (MD) is one such well-known disorder, associated with mutations that affect the protein components of muscular tissue. Certain chronic infections can lead to disorders associated with muscle inflammation, called myositis, usually because of autoimmune disease.

Myasthenia gravis. Among the most well-known neuromuscular junction disorders is *myasthenia gravis* (MG). This is an autoimmune disorder caused by antibodies that block acetylcholine receptors at muscles.

Amyotrophic lateral sclerosis (ALS) is a disorder in which either or both upper motor neurons from the cortex, or lower, alpha motor neurons in the spinal cord ventral root fail.

Infections. *Viruses* such as polio and rabies also attack motor neurons. Typically the viruses are taken up by the process of pinocytosis at motor neuron axon terminals and then destroy the motor neurons.

Diabetes. Diabetes, toxins, and infections can compromise and kill both motor neurons and sensory neurons upon which precise motor control depend.

Multiple sclerosis. *Multiple sclerosis* (MS) is a disorder in which the myelin coating around neural axons deteriorates over time. It is believed to be caused by an autoimmune process—when the body mounts an immune attack against its own tissue, in this case myelin. MS onset tends to be in the age range of 18 to 50. MS affects all neural axons in a highly variable manner that is not well understood. In the case of its effect on motor neurons the initial symptoms are motor weakness, often followed by partial or full paralysis of parts of the body. There can be periods of partial remission. There is no known highly effective treatment for MS.

Spinal cord injury

Spinal cord injury, mostly from traffic accidents, is one of the leading causes of motor dysfunction in the U.S. and world. In America, almost 18,000 new spinal cord injuries occur each year, 78 percent of which are in males.

Severe spinal cord injuries, by severing axons from the brainstem and motor cortex, disrupt both voluntary movement and control of organs and glands (such as the bladder) below the level of the injury.

A healthy nerve

A nerve affected by multiple sclerosis (MS)

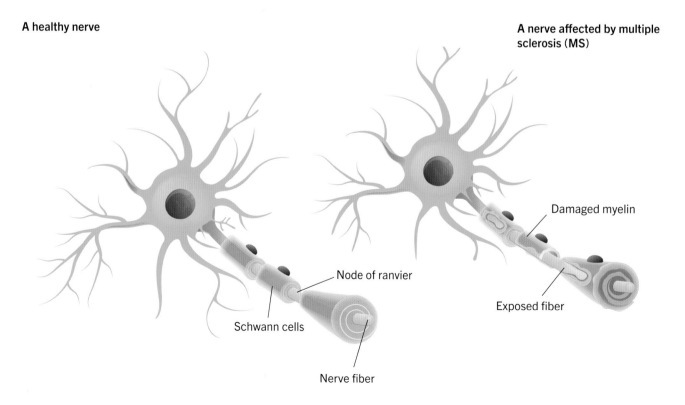

Damaged myelin

Node of ranvier

Exposed fiber

Schwann cells

Nerve fiber

Above: *Quadriplegia can be caused by spinal cord injuries.*

The part of the body paralyzed depends on the level of spinal segment damaged and the extent of the sensory or motor tract severed. Cervical damage can produce quadriplegia, loss of all motor control below the neck. Damage to lower-level segments, such as the thoracic or lumbar, may spare control of upper limbs but induce paralysis in the legs and cause loss of control of autonomic functions such as bladder control (paraplegia).

In the case of muscle control, attempts have been made to directly stimulate either the muscles themselves, or lower motor neurons that may still exist in spinal cord segments below the injury. This rehabilitation effort is based on the idea that signals to move muscles are still being generated in the cortical motor area, but cannot reach the muscles due to the interruption of spinal cord axon pathways. The strategy

REGENERATION

A fundamental mystery concerns the fact the mammalian central nervous system axons do not regenerate. If one cuts the optic nerve of a non-mammalian vertebrate such as a fish or frog, after several months the cells whose axons were cut regenerate the axons which regrow and innervate the original target, restoring function. Severed axons even regrow in the mammalian peripheral nervous system. A complete cut of the dorsal root ganglion nerve is followed by regrowth and re-innervation of its targets in a few months. But this does not occur in the mammalian central nervous system.

One difference in mammalian central versus peripheral nerve axons is that the myelination of central nervous system (CNS) axons is by oligodendrocytes, whereas peripheral axons are wrapped by Schwann cells. There has been some limited success with transplanting peripheral nerve Schwann cell "sheaths" into pathways to guide regrowth of CNS axons. This indicates that oligodendrocytes somehow actively block regeneration, whereas Schwann cells permit it. Much further research is necessary, however, to translate these results into effective therapies for spinal transections.

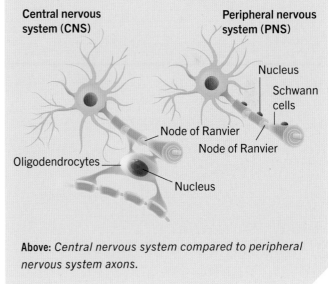

Above: *Central nervous system compared to peripheral nervous system axons.*

is therefore to record the cortical commands, bypass the compromised spinal cord segment electrically, and stimulate the muscles or motor neurons electrically.

A similar strategy is used for amputations. Here the lower motor neurons are intact, but there is no muscle to stimulate. Instead, sensors pick up the lower motor neurons signals and use these to operate a mechanical prosthesis, such as an artificial arm or leg. With practice the artificial limb can feel like a natural part of the person. Amputees often report

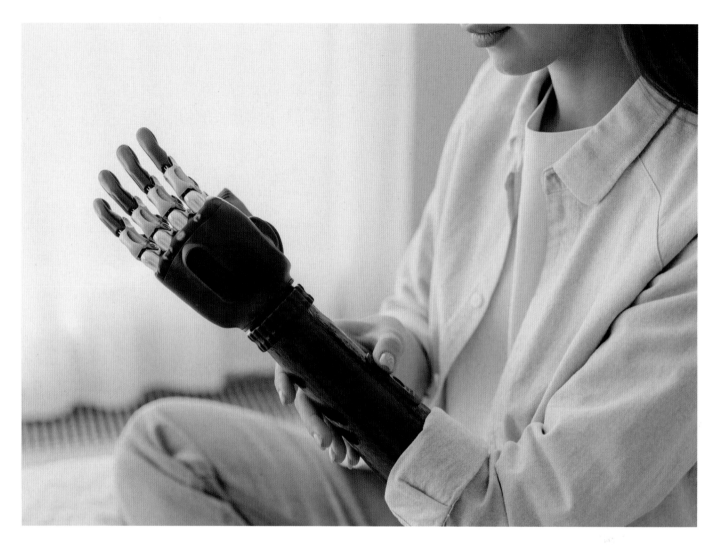

Above: *In prosthetic limbs, sensors pick up the signals from motor neurons.*

feeling sensation in a limb that is no longer there, possibly generated by activation of the skin whose representation in the somatosensory cortex is adjacent to the area that represented the missing limb, but now has no input from that limb.

Parkinson's and Huntington's diseases

There are two well-known neurological disorders associated with basal ganglia dysfunction: Parkinson's and Huntington's diseases. These disorders have quite different causes and produce nearly opposite movement deficiencies.

Normally the balance of the output of the basal ganglia to the thalamus projects enough excitation to select one movement alternative and inhibit all the others. In both Parkinson's and Huntington's diseases, the output from the basal ganglia is wrong, making the thalamic output from the cortex wrong, resulting in movement dysfunction.

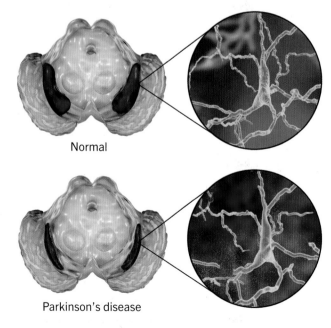

Normal

Parkinson's disease

Above: *The substantia nigra in Parkinson's disease.*

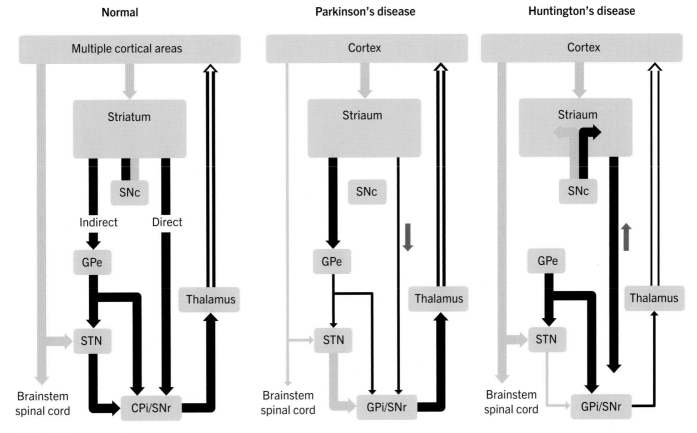

| Normal | Parkinson's disease | Huntington's disease |

Normal
- Multiple cortical areas
- Striatum
- SNc
- Indirect / Direct
- GPe
- STN
- Thalamus
- Brainstem spinal cord
- CPi/SNr

Parkinson's disease
- Cortex
- Striaum
- SNc
- GPe
- STN
- Thalamus
- Brainstem spinal cord
- GPi/SNr

Huntington's disease
- Cortex
- Striaum
- SNc
- GPe
- STN
- Thalamus
- Brainstem spinal cord
- GPi/SNr

Excitatory connections
Inhibitory connections

Parkinson's disease is caused by a loss of dopamine-producing neuron in the substantia nigra (SN), a midbrain structure connected to the basal ganglia globus pallidus. Inadequate dopamine input disinhibits the direct pathway output of the globus pallidus, leading to too much inhibition of the thalamus. It is then difficult to select any path through the frontal lobe to execute a movement plan. Parkinson's patients have difficulty initiating any movement sequence, although often, once moving, they can continue as long as there are no obstacles. One test for Parkinson's is to have the patient walk along a hall and then place a small wooden plank in their path, which would be easy for a normal person to simply step over. Parkinson's patients have trouble doing this: they cannot alter their movement due to this contingency because the basal ganglia are not functioning properly. The cause (or causes) of Parkinson's disease are not currently known. There appears to be some genetic susceptibility, and environmental toxins are suspected of acting on this genetic weakness in some people.

Huntington's disease is the result of a genetic mutation. The mutation causes reduced inhibition of the indirect pathway of the basal ganglia, resulting in excess excitation of the thalamus and motor output. Whereas Parkinson's

Above: *Operation of the basal ganglia in normal vs Parkinson's and Huntingdon's diseases.*

patients have difficulty initiating and modulating movement, Huntington's patients cannot stop moving.

Acute injuries

Acute traumatic brain injury (TBI) can occur from an impact to the head, violent shaking, or excessive rapid head movement such as whiplash in vehicle collisions, often called a concussion. In the U.S there are around 250,000 hospitalizations from TBI. Concussions can cause immediate brain damage, and longer lasting brain injury from swelling induced by the injury.

There has been an increasing awareness that traumatic brain injury can occur not only in obvious acute cases, but can develop over the long term from multiple head blows, each of which is not associated with obvious lasting dysfunction on its own. This awareness has brought about significant rule changes in sports such as American football, rugby, and even soccer to protect athletes from frequent blows to the head. One predictive sign for the possibility of brain damage from a head blow is the loss of consciousness, even if temporary.

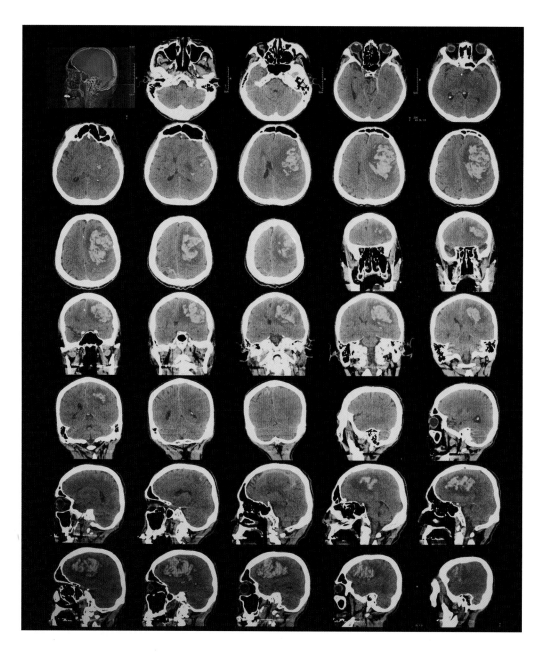

Right: *A CT scan showing a hemorrhagic stroke.*

More serious are penetrating brain injuries, involving direct damage to tissue, severe bleeding, fractures to the skull, and infection risk. Brain injuries can cause sensory, motor, or cognitive deficits.

Strokes

Strokes involve, at the minimum, an interruption to the blood supply to a particular part of the brain. There are two major types of strokes—ischemic and hemorrhagic.

Ischemic strokes involve the interruption of the blood supply to an area of the brain area, usually due to a blockage in the arteries, but without any rupture of the blood vessel or bleeding. When a rupture does occur, it is instead referred to as hemorrhagic stroke. Strokes can occur anywhere in the brain, and their effects on motor control depend on the location of the stroke. Strokes more often destroy neural cell bodies than axons of passage, although severe ones can destroy both.

Tumors

Tumors can have similar effects to those of strokes. The blood-brain barrier that tightly controls what can get from the vasculature into the brain also makes it difficult for immune cells to enter the brain to fight cancer. This makes the prognosis for brain neoplasms very poor, especially cancers of glia cells (gliomas).

Dystonia

Dystonia is a movement disorder in which muscles contract involuntarily in repetitive or twisting movements. Its effects can range from a small part of the body to the entire musculature. Its cause is unknown, and there is no cure, although some medications partially alleviate symptoms.

Dyskinesia

Dyskinesia involves involuntary, writhing movements of the face, arms, trunk, or legs. It is sometimes the result of levodopa medication used to treat Parkinson's disease, or from long-term use of some psychotic medications.

Ataxia

Any disorder that affects co-ordination, balance, or speech is called an ataxia. One of the most common causes of ataxia is cerebellum damage arising from TBI, strokes, or tumors.

Tremors

Tremors involve involuntary, rhythmic muscle contractions such as shaking. It can occur in any part of the body. Tremors have a variety of causes, such as side effects of some medicines, but some tremors occur during aging without any known underlying associated neurological disorder.

Tourette syndrome

Tourette syndrome (TS) is characterized by unwanted rapid, repetitive movements or vocalizations called tics. Its onset is in the age range of 2–15 and occurs about four times more often in males than females. There is no cure for TS, but treatments that reduce dopamine levels mitigate some symptoms. In some cases, Tourette symptoms lessen or even disappear with age.

Below: *A man walking with ataxia, from the Muybridge motion studies, 1887.*

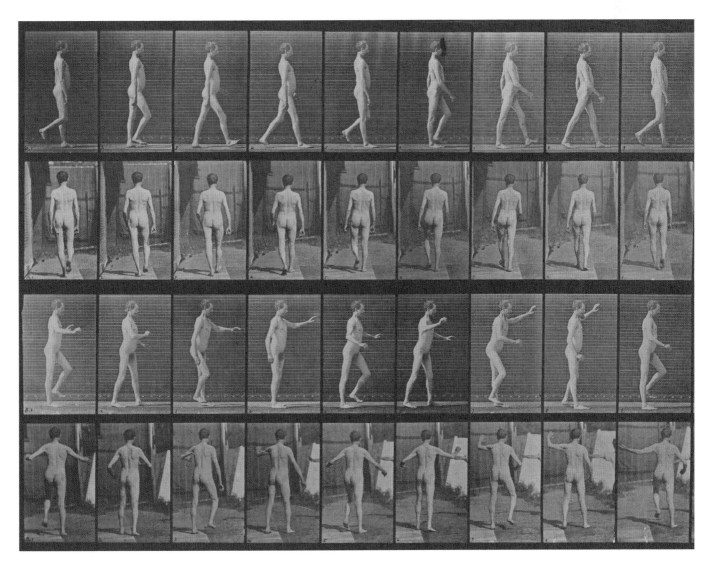

THE CONSCIOUS BRAIN

The vast majority of the activity of the nearly 100 billion neurons that make up our nervous system do not contribute to awareness or consciousness. Some of this activity, however, does influence emotions, and through this mechanism this activity reaches consciousness. In this chapter, we will examine the nature of consciousness, where it is located in the brain and its mechanisms in controlling our actions, as well as understanding the roles of intelligence and emotion.

Consciousness arises through the activity of the brain.

// Consciousness

We all tend to think that consciousness produces most of our behavior, but the truth is, we often simply react to external inputs because of our evolutionary and learned programming. Neural processing occurs with and without awareness. By most definitions of consciousness, most non-human animals have only unconscious neural activity. Large areas of the human nervous system that are shared with animals operate without consciousness. We have little insight into or understanding of how our brain allows us to hit a tennis ball or speak a grammatical sentence. In language, particularly, we are conscious only of the content of our

Below: *Consciousness can be understood as "scaffolding", as explained in a Steven Peterson's and Marcus Raichle's theory.*

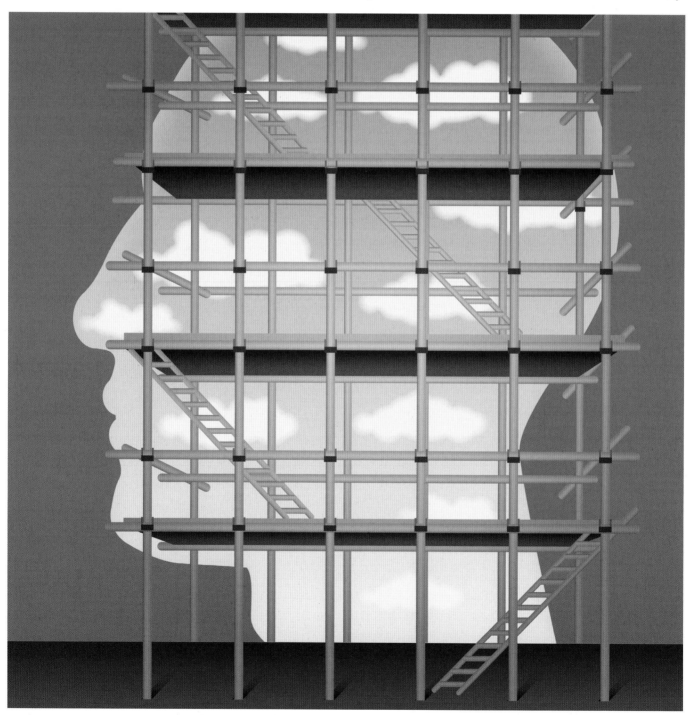

thoughts, but not of what generated the words and grammar of the thoughts.

Many brain scientists suggest that below consciousness is a level called awareness that many animals possess. Awareness can be thought of as related to the difference between perception and sensation. For example, virtually all animals sense contact with their skin, which is a sensation that can trigger some behavior, like avoidance. But suppose a grasshopper lands on the back of your family dog. He turns to bite at the irritant, but upon seeing it, "perceives" that it is also something he can eat, having eaten grasshoppers before. A dog has, through memory and experience, some "concept" of a grasshopper, which he now perceives as being what is on his back. If a bunch of grasshoppers were hopping around on your lawn, you could probably train your dog to fetch one.

In this scheme, the difference between awareness and consciousness is primarily associated with language. Humans, with language, can generate an infinite number of concepts, about grasshoppers, expressible in sentences, including their appearance, properties, behavior, that they are insects, and so forth. Clearly, by this understanding, some non-human animals may have awareness but do not possess language-based consciousness.

Theories and experiments about consciousness

Consciousness as a temporary "scaffolding"

Motor learning, like learning to ride a bicycle, is an example of a procedure in which we start being very conscious of the components or steps during its execution, but then lose consciousness of all the steps upon attainment of expertise. In a theory called the scaffolding framework, developed by Steven Peterson and Marcus Raichle, we use conscious processing during practice while developing skills (or memories) because consciousness is the scaffolding that allows execution of a novel sequence before the skill is well developed. Once the task is learned, the brain areas used for the task change. Earlier, we saw that the initial execution of a new task used the premotor cortex (PMC) areas of the brain that received sensory input for conscious feedback. But, once the task was learned, execution control switches to the supplementary motor area (SMA), which can carry out the activity without requiring conscious access to lower-level components (see pages 121–3). In effect, the scaffolding support structures are removed once permanent neural structures have "stored" the sequence.

The Libet "free will" experiment

In the late 20th century, a neuroscientist named Benjamin Libet conducted an experiment that generated fundamental

Above: *Benjamin Libet.*

questions about the nature of free will. In this experiment, Libet recorded the brain EEGs of a subject who was told to press a button voluntarily at any time the subject decided to do so. The subject also viewed a special clock by which he could report the time of his decision to push the button. Sitting in the next room monitoring the brain electrical potentials from the supplementary motor area, Libet found that these potentials indicated that the subject was going to press the button about half a second before the subject "decided" to do so. The major question that arose was whether, given these data, conscious decisions are causal. That is:

1. Does a person decide to do something, and then the brain activity follows,

or

2. Is what the subject believes to be the decision merely an epiphenomenon of unconscious brain activity?

Many neuroscientists have suggested that the Libet experiment supports the second alternative. Interestingly, Libet himself believed that the "urge" to push the button was generated without explicit consciousness, but that people have a conscious "veto" power over actually taking the action of pushing the button, or any other action.

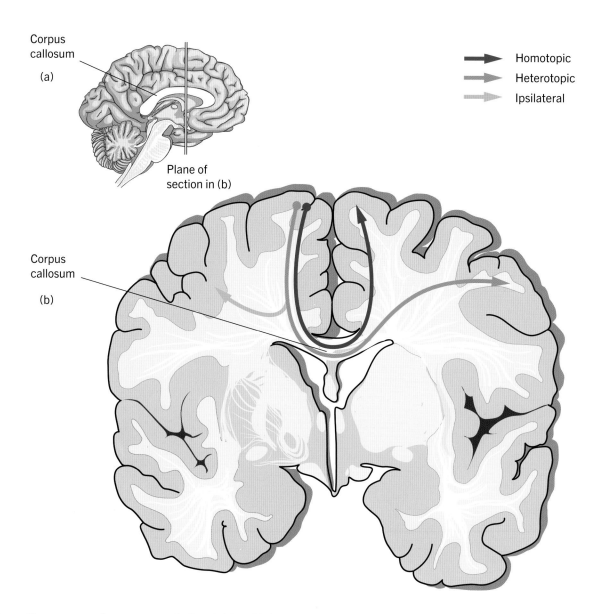

Corpus
callosum

(a)

Plane of
section in (b)

Corpus
callosum

(b)

Homotopic
Heterotopic
Ipsilateral

Above: *The corpus callosum connects the right and left hemispheres of the brain. Language is usually located in the left hemisphere, and it is through this fiber tract that it can be accessed by the right hemisphere.*

The Gazzaniga "interpreter" in the left hemisphere

The language ability is located in the human brain's left hemisphere (in most people).

Normally, information in the right brain hemisphere has access to neural language mechanisms via the fiber tract connecting the two hemispheres called the corpus callosum. This pathway of over 200 million axons is the largest neural tract in the brain.

Most of the connections traversing the corpus callosum are either:

1. "homotypic" connections to and from the same areas on each side of the brain, or
2. "heterotypic" connections from one side of the brain

to a higher order area on the opposite side of the brain.

We know we have linguistic consciousness of activity in the left brain because this activity can directly drive language understanding and production areas. We have linguistic consciousness of neural activity on the right side of the brain because this activity is relayed across the corpus callosum to activate similar circuits on the left side that have access to language.

A fundamental question that neuroscientists have long asked is "What awareness would we have of activity in the right side of the brain if it were not connected to the left side by the corpus callosum?" The answer was provided by research on so-called split-brain patients whose corpus callosums were severed to block seizures from traveling from one side of the brain to the other. In this research by Michael Gazzaniga and colleagues, the subject was asked to focus on a central dot. Images were flashed in either the left or right visual field for long enough to reach awareness, but not long enough to permit an eye movement.

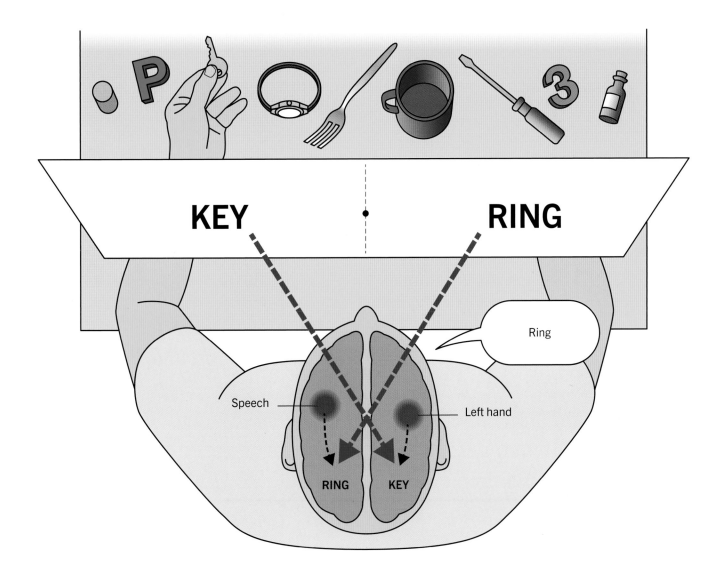

Above: *Gazzaniga's flashed word experiment for split-brain patients.*

Split-brain patients were able to report the name of the object flashed in their right visual fields (that were processed by the left, linguistic side of the brain), like normal subjects. But they had no ability to verbally report any knowledge about objects presented in their left visual fields, that is, those processed by the right side of the brain. That is, they had no conscious awareness of left visual field objects. However, if the split-brain patients were asked to reach with their left hands (controlled by the right side of the brain) to the flashed object, they did so reliably, while still denying knowing anything about the visual presentation of the word for the object. Thus, the disconnected right side of the brain could generate behavior based on information it received, but this information could not be consciously accessed or reported.

In a similar experiment, Gazzaniga and colleagues presented a single command word in the left visual field of split-brain patients. Although the patients had no awareness of the word, they often acted out what it commanded. Then, when asked to verbally report why they were behaving as they did, the patients typically confabulated a story consistent with their behavior, still without knowledge of the word that had been flashed.

Some neuroscientists and philosophers interpret this experiment, like the Libet experiment described earlier, as showing that conscious free will is not causal. That is, it does not precede and bring about behavior, but rather is part of an ongoing dialog by the left linguistic side of the brain to make sense of what is going on in the world. Another example of the ability of the brain to use information that it does not have conscious access to is blindsight (see page 89), when people are unaware of visual stimuli, but can still react to those stimuli.

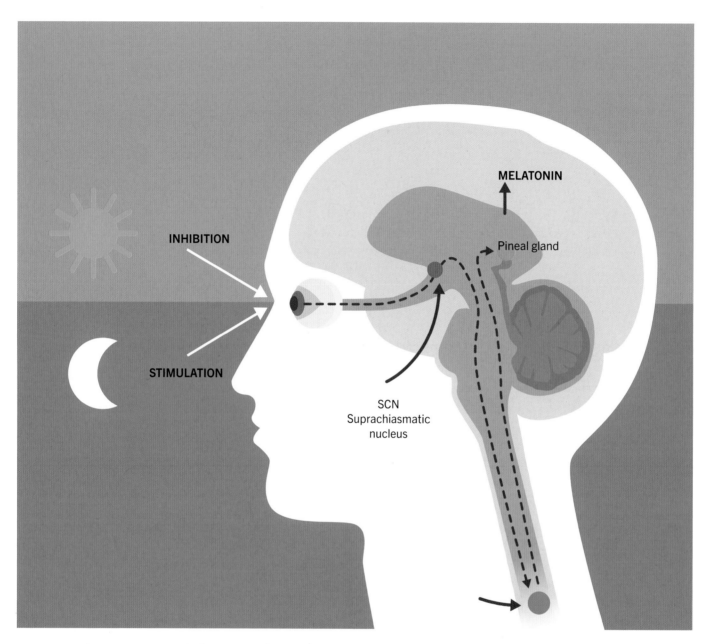

MELATONIN

Pineal gland

INHIBITION

STIMULATION

SCN
Suprachiasmatic
nucleus

Sleep

Sleep is a transition from consciousness to unconsciousness. The brain is still active during sleep, however. In fact, brain activity during the REM (rapid eye movement) phase of sleep is difficult to distinguish from that of the awake brain. The transition from being awake to asleep has been studied to explain what parts of the brain, or activity in the brain, specifically underlie consciousness. Moreover, all vertebrates, including even fish, sleep.

One of the first things sleep researchers noticed about stages of sleep was the existence of a stage associated with rapid eye movements, called REM sleep. This is the stage of sleep when most dreaming occurs. The other stages of sleep are called non-REM and are numbered N1–N3 (or N4). The

Above: *The circadian rhythm forms our internal body clock, with light providing the cues for the sleep–wake cycle.*

various sleep stages tend to occur in either ascending or descending order.

Dreams can include the experience of moving or carrying out motor actions. During such dreams, prefrontal and frontal brain areas are activated, but the final output of primary motor cortex to physically generate the actions is suppressed (usually). Sleep is important in consolidating learning, and long-term sleep deprivation induces psychoses. However, many important aspects of the function of sleep remain unknown.

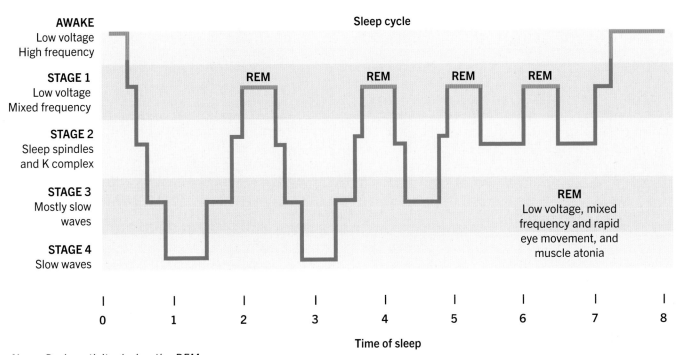

AWAKE
Low voltage
High frequency

STAGE 1
Low voltage
Mixed frequency

STAGE 2
Sleep spindles
and K complex

STAGE 3
Mostly slow
waves

STAGE 4
Slow waves

Sleep cycle

REM   REM   REM   REM

REM
Low voltage, mixed
frequency and rapid
eye movement, and
muscle atonia

0 1 2 3 4 5 6 7 8

Time of sleep

Above: *Brain activity during the REM stage of sleep closely resembles that of the awake brain.*

Below: *The sleep-wake cycle shows that there are different periods of the day for our peak activity in different areas.*

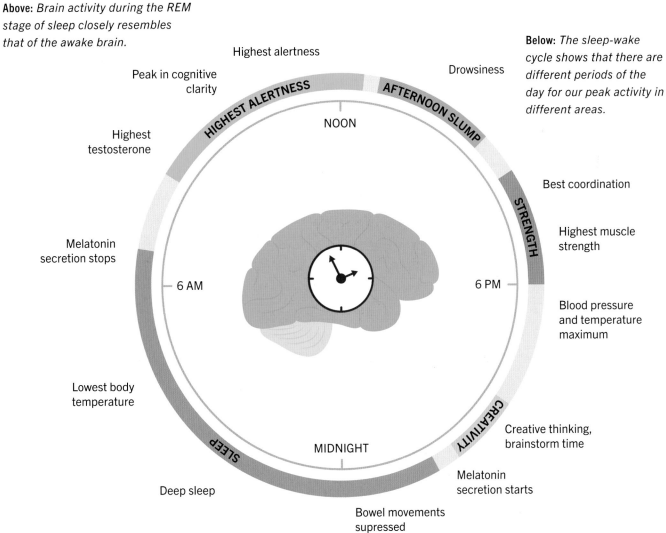

Highest alertness

Peak in cognitive clarity

Drowsiness

HIGHEST ALERTNESS

AFTERNOON SLUMP

Highest testosterone

NOON

Best coordination

STRENGTH

Highest muscle strength

Melatonin secretion stops

6 AM

6 PM

Blood pressure and temperature maximum

Lowest body temperature

CREATIVITY

SLEEP

Creative thinking, brainstorm time

MIDNIGHT

Melatonin secretion starts

Deep sleep

Bowel movements supressed

// Emotion

Emotions are brain states that arise either from external sensory inputs or from internally generated neural activity. Emotional states influence behavioral choices and the interpretation of incoming sensory data. Although the autonomic nervous system consists of two opposing subdivisions (sympathetic versus parasympathetic), there are clearly more than two distinct emotions. But how many? A central question in emotion research is whether the vast range of emotional experience is produced from a finite set of "fundamental emotions," with the experienced emotion being a sort of algebraic sum of several fundamental ones.

Models of emotion

Most current theories of emotion suggest that there are several emotional "axes" with opposite emotions at each axis endpoint, and a continuum of emotional experience along the axis. For example, happiness appears to the opposite of sadness, but one can be very happy, very sad, or at some neutral point between these.

One approach to emotional bases suggests the two axes of:

1. Valence—pleasant versus unpleasant.
2. Arousal—the intensity of the emotion.

The valence axis can be interpreted several ways. Richard Davidson proposed a fundamental behavioral dichotomy of approach versus withdrawal exhibited by all mobile animals. Animals approach pleasant things like food and withdraw from unpleasant ones like extreme temperatures or predators.

When Charles Darwin made his famous voyage on the Beagle, as well as developing his theory of evolution, he observed that various cultures he encountered all seemed to exhibit similar fundamental emotions. His ideas were published in The Expression of the Emotions in Man and Animals (1872). This idea was quantified in more recent times by the research of Paul Ekman, who came to a similar conclusion. Ekman postulated six basic emotion types that could be identified by associated facial expressions. The six emotions are anger, happiness, disgust, surprise, sadness, and fear.

These expressions exist on continuum axes with opposite endpoints, like happiness and sadness, and there is a continuum of intensity for all six emotions such that any given emotional state can be considered to be a point in the six-dimensional emotion space.

Neural substrates of emotion

Emotional experience affects virtually the entire brain, but some particular emotions or emotional axes can be associated with specific neural substrates. The emotional axis of approach/withdrawal involves processing within the medial regions of the prefrontal cortex in the right and left cerebral hemispheres. The two cerebral hemispheres appear to be associated with opposite ends of this continuum. Specifically, left hemisphere processing is biased to promote approach behaviors, while right hemisphere processing is biased to promote withdrawal behaviors.

This dichotomy is seen not only in changes in activity in the two hemispheres associated with these emotional states, but also with brain damage. Damage to the left frontal lobe that reduces its approach influence can result in severe depression, a state in which the primary symptom is withdrawal and inactivity. Damage to the right-hemisphere

Left: *Paul Ekman.*

Right: *The facial expressions for the six basic emotions.*

Sadness

Disgust

Anger

Surprise

Happiness

Fear

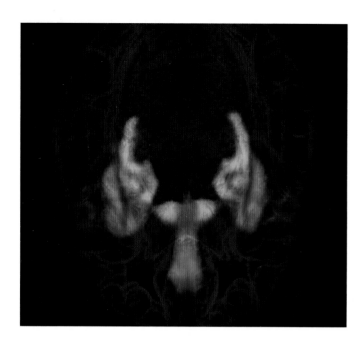

Above: *A PET scan highlighting limbic areas.*

system reduces its withdrawal function, biasing the patient to be socially engaging even when such behaviors are no longer appropriate. Some patients with right frontal damage appear manic.

The limbic system

The limbic system (see page 53) is the major seat of neural processing for emotion. This is particularly the case for learned rather than instinctual emotional responses. The earliest mammals added this system to the brainstem and cerebellum of primitive, non-mammalian vertebrates. Two structures in the limbic system are particularly important for learned emotional responses:

1. The hippocampus.
2. The amygdala.

The hippocampus has reciprocal connections to the entire sensory cortex (temporal, parietal, and occipital lobes), as well as to the decision-making areas of the frontal lobe

The amygdala, just in front of the hippocampus, has a functional relationship with the ventromedial, prefrontal, and orbitofrontal cortices. These areas operate with the amygdala to learn and mediate our ability to respond and act in:

1. Dangerous situations.
2. Socially complex situations.

Our consciousness is informed of the assessment of danger in the above situations via emotions.

Low and high roads

The amygdala is part of a behavior circuit that mediates rapid responses to learned danger. One way it does this is by short-circuiting the time-consuming, high-resolution processing of sensory input by visual cortex. The short stimulus-to-response pathway has been called the low road by neuroscientist Joseph LeDoux.

The high road for visual input is the normal sequence of signal processing from the retina to the thalamus to multiple cortical levels to identify objects in the scene. Unfortunately, this pathway takes time. If you see a dark, linear object just in front of you that might be a snake, you must act quickly. The low road involves a projection to the thalamus that then directly projects to the amygdala, which projects to the frontal lobe, which generates behavior. This allows you to

Below: *LeDoux's high and low road processes.*

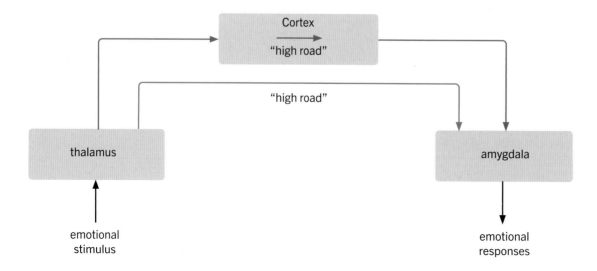

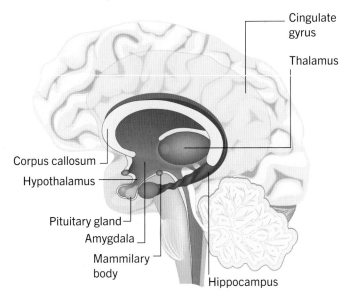

Cingulate gyrus

Thalamus

Corpus callosum

Hypothalamus

Pituitary gland

Amygdala

Mammilary body

Hippocampus

jump out of the way of the snakelike thing in front of you before the cortical circuitry has completely identified the object, and before this even reaches consciousness. Of course, you may jump back from what turns out to be a stick on the ground, as you later identify it through the high road, but the fast reaction is generally adaptive for survival.

The amygdala and fear learning

Damage to the amygdala does not usually impair emotional responses to innately aversive or rewarding stimuli, but rather to emotional responses that have been learned, such as that long thin black objects on the ground are sometimes snakes. This function of the amygdala has been investigated using a paradigm called fear conditioning. In fear conditioning an initially neutral stimulus acquires aversive properties (becomes conditioned) by being paired with an aversive event. This is a form of classical conditioning (see pages 149–50).

In the social domain, the human amygdala is particularly important for identification and responses to fearful facial expressions. Patients with amygdala damage are particularly impaired at evaluating fearful facial expressions, whereas their ability to evaluate all other facial expressions are relatively normal. fMRI imaging has shown that normal subjects do not have to be aware of seeing a fearful face for the amygdala to respond. Fearful facial expressions flashed quickly (subliminally, as in priming, see page 150), still activate the amygdala as strongly as those that the subject is aware of seeing.

Right: *The location of the amygdala.*

The orbitofrontal and ventromedial prefrontal cortex

The amygdala is reciprocally connected with two cortical areas, the ventromedial prefrontal and orbitofrontal cortical areas. Together with the amygdala, these areas function to create memory contingencies whose details are stored in these areas. These cortical areas are particularly important in the regulation of social and emotional decision making, that is, our ability to evaluate social information and, through emotion, act on it, usually by inhibiting some impulse, like joking about the tie your boss is wearing. Our behavior thus reflects the combined influences of our personal desires and impulses, versus social constraints.

The frontal lobes play a critical role in selection using sensory information. In particular, the dorsolateral prefrontal cortex is important for selecting information through its holding of working memory. But, choosing actions also requires using information about values, emotional state, and the social situation. Consider a driver tempted to make a dangerous lane change in heavy traffic. This temptation is restrained by the fear of an accident, potential traffic citations, and the honking horns of other drivers. The ventral

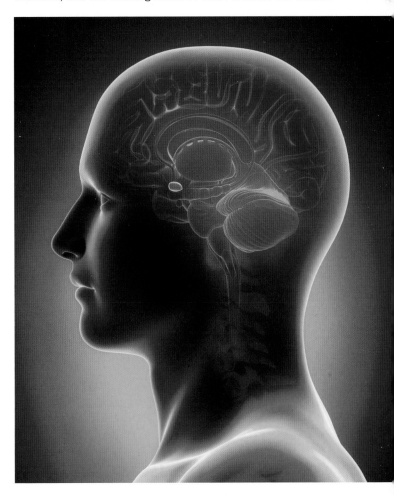

Below: *Ventral prefrontal areas.*

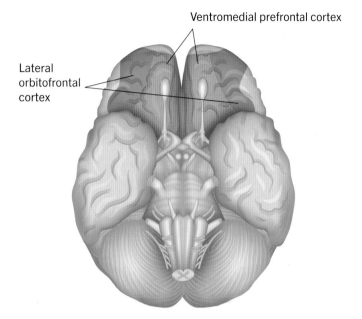

Ventromedial prefrontal cortex

Lateral orbitofrontal cortex

Left: *Antonio Damasio.*

prefrontal areas, in association with the amygdala, are essential for the selection of actions based on evaluating decisions within a social context.

Patients with damage to the ventral prefrontal cortex have deficits in social decision making. Their responses are overly triggered by sensory input, ignoring important social cues. These patients are insensitive to social norms, and they often have difficulty inhibiting inappropriate social responses, such as aggressive impulses. Patients with ventral prefrontal or orbitofrontal lesions are more prone to exhibit antisocial behaviors, such as stealing or violent outbursts. Some of these patients have what is called acquired sociopathy because of their tendency toward violence and lack of concern for social consequences. Brain imaging studies indicate reduced glucose metabolism in the orbitofrontal cortex in such individuals.

Gut feelings

Activity in ventral prefrontal cortex activates, via the autonomic nervous system, a "gut feeling" emotion about the risk of performing a dangerous or socially unacceptable action. Antonio Damasio referred to this gut feeling mechanism as the somatic marker. Somatic markers are bodily sensations derived from ventral prefrontal physiology where similar events experienced in the past became associated with negative or positive emotions through learning involving the amygdala. These somatic markers anticipate the affective consequences of actions, without conscious recall of the specific similar situations. Even when representations of previous similar incidents arise in working memory, if the ventral prefrontal areas are damaged, these remembrances are stripped of their emotional content and do not produce the correct brake on inappropriate behavior.

Risk and reward

A gambling experiment illustrated the improper assessment of risk in ventral prefrontal damage patients. In this experiment subjects could select from one of two decks of cards and learned through trial and error the payoffs connected with each deck. The cards in one deck often provided large payoffs, but also even larger losses such that the average payout was negative. The rewards and penalties of the other deck were smaller but had a small net favorable payout. Control subjects always eventually learned to choose from the milder, better payoff deck, but patients with ventral prefrontal lesions favored the riskier decks because they were attracted to the frequent high payouts even though they eventually would be offset by a significant loss.

Above: *A gambling experiment showed how ventral prefrontal damage led to inaccurate risk assessments by patients.*

// Intelligence

Intelligence is a difficult concept to define, and even more difficult to relate directly to the structure and activity of the nervous system. Most neuroscientists generally agree that vertebrates are generally "smarter" than invertebrates, mammals smarter than non-mammalian vertebrates, primates smarter than mammals, and humans smarter than other primates. There is a general increase in brain size throughout this spectrum. Invertebrate cephalopods, such as the octopus, however, can produce extraordinarily complex behaviors using fairly

Below: *A vocabulary aptitude test from 1942.*

small central nervous systems. In humans, there have been very intelligent people with very small brains, and some pathological cases of people with normal or above normal intelligence who have almost no neocortex.

In humans, the idea of a one-dimensional "intelligence quotient" (IQ) arose from psychometric measures of a high correlation in human performance on a variety of tasks considered to require intelligence. This was mapped to a scale in which a score of 100 represented some "average" human intelligence, and scores above and below were so many statistical standard deviations from that average.

Divisions of intelligence

One of the first instances where the high correlation across different types of tasks sometimes breaks down is in verbal versus spatial intelligence. Although performance on these two different types of task tends to be correlated in most people, there are significant numbers of people who excel at verbal, but are poor at spatial tasks, and vice versa. This led to partitioning of IQ and similar tests such as the SAT into verbal versus spatial/mathematical sections, with separate scores for each. This distinction was found to have some neural basis in the specialization of the left half of the brain for linguistic tasks, versus the right for spatial ones.

Some researchers, such as Howard Gardner, have argued for further divisions of intelligence. Gardner argued for seven major types of intelligence:

1. Verbal-Linguistic—words and language.
2. Logical-Mathematical—logic and mathematics.
3. Musical—melodies, rhythm, pitch, tones.
4. Bodily-Kinesthetic—body movement expertise.
5. Spatial-Visual—layout of objects in three dimensions.
6. Interpersonal—social intelligence.
7. Intrapersonal—knowledge of one's own capabilities, tendencies.

More recently Gardner and others added an eighth intelligence called naturalistic associated with knowledge of the biological environment.

Verbal-linguistic and the combination of logical-mathematical and spatial-visual roughly correspond to the previous intelligence segmentation along linguistic versus spatial/mathematical lines. The other categories reflect negative correlations found in some of the population where high scores on one or more of these divisions does not correlate with high scores on the others. Regardless of such findings, there is little known neuroscientific basis for these additional intelligence subdivisions. It is also unclear the extent to which some of these "intelligences" reflect exposure to specific learning/training rather than being based on an underlying innate capability.

Nature versus nurture

The oldest debate in the history of mental science is about the role of innate capacities of the mind versus capabilities that have been learned. Language, for example, is developed by all normal human children exposed to it, but feral (wild-raised) children have been found throughout history who lack language, and cannot learn it after adolescence. People who have grown up with extreme optical problems in both eyes, such as congenital cataracts, are not able to achieve normal

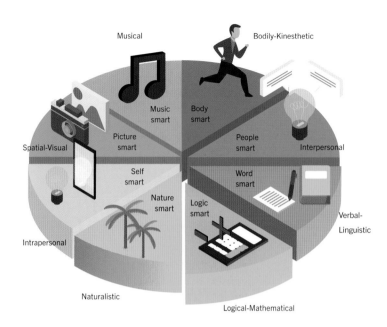

Above: *The eight types of intelligence.*

pattern vision as adults even if the optical problems are surgically corrected.

The brain's structure is not simply a product of a genetic blueprint. The human genome has about 20,000 protein-coding genes, but there are 86 billion neurons in the brain, and over a quadrillion synapses. During late development, synapses and neurons are removed by pruning that have not been sufficiently activated by either other inputs or other neurons in the brain. There are "critical periods" in development during which insufficient activation of some neural pathways causes the deletion of that pathway, which cannot be reversed by later activation. On the other hand, it is certainly the case that genetic coding controls the susceptibility to critical periods or the ability to learn in certain domains.

Without any sound neuroanatomical or neurophysiological basis for disentangling nature versus nurture in intelligence, or the relationship between scores on intelligence tests and performance, the usefulness of testing for intelligence becomes an empirical matter. An argument taking place in many college science graduate programs is whether standardized graduate aptitude testing should be dropped because undergraduate grades are a better predictor of graduate grades than standardized tests. However, this argument ignores the obvious question of whether graduate grades predict post-graduate science performance. Perhaps good grades reflect students' inclination to do as they are told, not highly motivated and original thinking that are known to be attributes of outstanding scientists.

// Learning

Learning is the process by which *memory* is created. Generally, it is the result of some experience such as sensory input or task performance, often strengthened by repetitions of the experience. Memory is the change in the nervous system resulting from learning. The most common form of memory is a change in the synaptic strength of synapses in sensory processing, motor output, or associative brain circuits. Memory changes can also occur by addition or deletion of synapses, or even entire neurons.

Learning shares many characteristics with *development*. During development, neurons and synapses are added to the nervous system, followed by a pruning stage in which many synapses and neurons are eliminated. The mechanisms by which these changes occur in learning and development are intensely being researched by neuroscientists.

Declarative and non-declarative memories

Memory formation can occur with or without conscious intention to remember something, or with or without conscious awareness of the memory formation. These are called declarative versus non-declarative memories. A typical example of creation of a declarative memory is learning the capitals of all the countries of the world by verbally repeating and rehearsing the list of country names and capitals. While doing this, you are explicitly aware of

Below: *Learning to ride a bicycle is an example of non-declarative memory.*

the process and you can verbally report the results of the memorization.

A typical example of non-declarative memory is learning to ride a bicycle. At first one is consciously aware of the required activities of pedaling, steering, and balancing. After practice one tends to focus on the road ahead and not these procedural elements, and people generally are unable to explain how it came to be that they mastered bicycle riding.

There are four main types of non-declarative memory:
1. Non-associative learning.
2. Conditioning.
3. Procedural (motor) learning.
4. Priming (modulation of the perceptual representation system).

Non-associative learning: habituation and sensitization

Two forms of non-associative learning, habituation and sensitization, occur in all animals, from primitive invertebrates to humans, and typically are mediated by sensory neurons or neurons closely connected to sensory neurons. They are called non-associative learning because they typically involve modulation of a single sensory/motor/perceptual pathway rather than requiring the formation of an association between several pathways.

Habituation occurs when repeated stimulation leads to a decline in the perception or response to the stimulation.

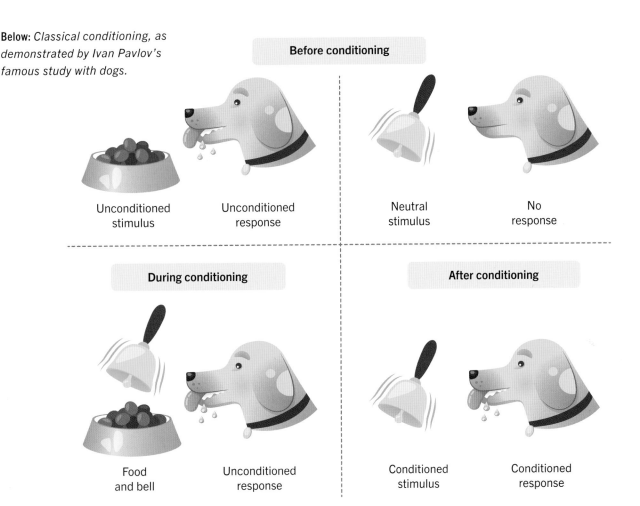

Below: *Classical conditioning, as demonstrated by Ivan Pavlov's famous study with dogs.*

Before conditioning

Unconditioned stimulus   Unconditioned response

Neutral stimulus   No response

During conditioning

Food and bell   Unconditioned response

After conditioning

Conditioned stimulus   Conditioned response

For example, when entering a room with a fan making some noise, you might initially be quite bothered by it, but after a while become totally unaware of the same noise. Habituation often occurs in sensory receptors themselves because many receptors for environmental stimuli are transient or rapidly adapting, generating fewer spikes to a constant or repeated stimulus than when the stimulus is initially applied.

Sensitization is in some sense the opposite of habituation. In sensitization, repetition of the stimulus leads to an enhanced response to later versus earlier applications. A common example of sensitization is the experience of being outside on a warm summer evening and occasionally feeling a light touch on one's leg. Then, at some point, one receives a mosquito bite on the leg, after which the previously hardly noticeable leg touches produce swatting responses. The synaptic mechanisms of sensitization involving the gill withdrawal reflex of the marine invertebrate Aplysia (sea slug) were explained by Eric Kandel, for which he was awarded the Nobel Prize in Physiology or Medicine.

Conditioning

Conditioning is another form of non-declarative learning that occurs across the animal kingdom. There are two major types: classical (Pavlovian) and operant.

Classical conditioning was famously studied by Ivan Pavlov, and is often called Pavlovian conditioning. Classical conditioning involves four elements, two of which are stimuli, and two responses:

1. The *unconditioned stimulus* is a stimulus that naturally produces a response in some animal. This can be an attractive (appetitive) stimulus such as the sight of food, or an aversive stimulus, such as a painful shock.
2. The *unconditioned response* is the natural response of the animal to the unconditioned stimulus. For an attractive or appetitive stimulus such as the sight of food, an unconditioned response would be salivating. The unconditioned response to a painful shock might be moving away from the source of the shock.

3. The *conditioned stimulus* is a stimulus that is normally neutral, such as a weak light flash that evokes no particular response from an animal.
4. The *conditioned response* is the result of classical conditioning in which the prior neutral stimulus, by being paired with the unconditioned stimulus, produces the conditioned response.

Operant conditioning is the generation of a behavior or behavior chain in response to a stimulus. Operant conditioning was studied by B. F. Skinner in pigeons, among other animals, using a procedure called *shaping*. In this experiment, the experimenter put a pigeon in a chamber (called a "Skinner box") in which there was a disk the experimenter wanted the pigeon to learn to peck, and a food pellet delivery chute. At first, as the pigeon moved around the box randomly, a food pellet might be delivered whenever the pigeon got close to the disk to be pecked. After this behavior was acquired, the food pellet might be delivered only when the pigeon pecked the wall on which the disk was mounted. After this, the food delivery required the pigeon to peck the disk. Shaping could also be done with aversive stimuli that caused the animal to avoid doing certain things.

Procedural (motor) learning

Learning to ride a bicycle is a form of non-declarative learning called procedural or motor learning. Nearly all animals will improve their performance on a motor task with practice. Animals do not have any language capability to explain how they improved on the practiced task (non-declarative), but even humans, despite their language, are usually unable to explain verbally how they hit a tennis ball more accurately or parallel-park. One mechanism of procedural learning is "chaining," where each step in a motor sequence automatically activates the next step upon its completion.

Above: *B. F. Skinner inserts a pigeon into a chamber.*

Priming (modifying the perceptual representation system)

The concept of *priming* is similar to what used to be called *subliminal perception*. Subliminal perception has become a contentious issue in mainstream psychology. A famous anecdotal instance of subliminal perception was the alleged practice of inserting, at regular intervals, a single frame showing an appetizing picture of a box of popcorn in movie films. These single image frames were not *consciously* perceived by the viewers. The intention was, of course, to induce movie patrons to buy popcorn at the theater or drive-in. This practice quite naturally generated opposition among the public and was legally banned. Later studies indicated that the practice was not actually effective in selling popcorn. The results of these studies were generalized in many textbooks to the conclusion that subliminal perception does not exist.

This is not quite true, however. The results of experiments like that of Anthony Marcel have been extensively replicated and are commonly used in neurological assessment for brain damage. In this experiment, an image of a word is first briefly flashed like the popcorn image in the movie scene, but only a single time. A noise mask is then presented to destroy the iconic image of the word. Then a string of letters is presented that may or may not be a word, and, if a word, it may or not be related to the initially flashed but not consciously perceived word.

The robust result of this and similar procedures is that the reaction time to indicate whether the last string of letters is a word is significantly faster if it was primed by the initial flashed word being related to it. Because this priming occurs without any conscious awareness of the initially flashed word, it is a kind of non-declarative memory that is now called priming.

Above: *Subliminal advertising involved inserting a frame that was not consciously seen by viewers, but it was a less effective practice than once believed.*

Below: *The procedure of priming.*

(a)

Bread or

Truck

Priming word presented in short flash that subject does not consciously perceive. Word may or may not be related to later target word.

(b)

XXXXX

Noise "mask" presented to destroy iconic image.

(c)

Sandwich

Reaction time to presentation of possible word significantly faster if related to prime word than not.

// Sensory and short-term memory

Different types of memory operate on different time scales with different capacities, mechanisms, and brain loci. There are three main divisions of memory by time scale:

1. Sensory (iconic) memory is a literal image that lasts only for a few seconds.
2. Short-term memory lasts seconds to minutes and depends on rehearsal.
3. Long-term memory lasts a lifetime.

Sensory memory

Sensory memory has been most extensively studied in the visual and auditory domains, although it exists for all senses. Visual sensory memory is called iconic, whereas that for auditory memory is called *echoic*. The key features of sensory memory is that it:

1. Decays swiftly, in seconds.
2. Is a literal sensory, not *semantic* representation of the input.
3. Has a larger capacity than the next, short-term, memory stage.

Iconic visual memory is a photograph-like mental image of a visual input that is accessible for only a few seconds. Take the following experiment. An image of 12 letters in a 3 row by 4 column array is flashed briefly (50 msec). A second or two later, a high, middle, or low auditory tone indicates which row is to be reported (middle panel). The person then verbally reports the letters in this row (bottom panel). Most people are able to report the row cued accurately.

An important aspect of this experiment is that humans generally have a short-term memory limitation of about seven items. Despite this short-term memory limitation, for a few seconds, all 12 flashed letters are accessible, because the person can report any row of 4 letters out of the 12. This experience is reported as though one were "reading" the letters off of an internal image in the mind of the flashed array. This internal image decays rapidly, however, and is gone after a few seconds, like the afterimage produced by staring at a light bulb.

A phenomenon similar to visual iconic memory in the auditory system is called echoic memory. Echoic memory lasts on the order of ten seconds, but still much less than the duration of auditory short-term memory. A familiar instance of echoic memory is hearing a flight announcement at an airport. For about ten seconds one is not only aware

of the information, but one can hear in the mind's "ear" the actual accent of the person, noise, speaker clicks, and other literal aspects of the announcement. A minute later, one is aware of the announcement information, but much of the literal sound memory is gone.

A common property of both iconic and echoic sensory memories is their literalness. One is not conscious of their semantic content until the information is "read out" of the sensory buffer and analyzed by perceptual recognition circuitry of the brain and sent to short-term memory.

Short-term memory

In first half of the 20th century, a common method of making phone calls was to call the operator to get the person's phone number, hang up, and then dial the number from memory. The phone company tasked psychology professor George Miller with finding out how many digits people could remember for the time needed to make a call this way. He found that the number was 7 +/-2, that is, most people could deal with 7 digits, with some able to retain less and some more than this. This is called the "span of immediate memory," or the "span of short-term memory."

Short-term memory is different from sensory memory because it:

1. Is semantic based.
2. Depends on rehearsal.

Short-term memory is semantic based because the rehearsed information must first be converted into a meaningful semantic identity before rehearsal. For example, the letters in a visual image must be recognized and rehearsed by sub-vocally repeating the sound of each in a repeating sequence. Even in the case of hearing a string of letters they must be correctly recognized to be rehearsed. If a letter is misrecognized, the wrong letter is what will be rehearsed.

Short-term memory appears as though it were a mechanism with about seven "slots" that are loaded from the rapidly decaying sensory memory. In the time it takes to load the seven slots, the sensory memory has decayed and no additional items from that source can be retained. The seven or so items that do become loaded, however, can be retained indefinitely by rehearsal. The reason that short-term memory is described as having a duration of minutes rather than longer is that rehearsal of short-term memory contents begins the process of converting short-term to

Below: *The iconic memory experiment.*

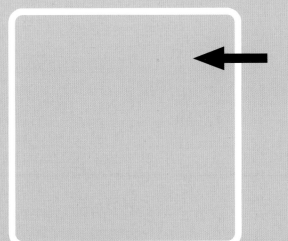

Array flashed for 50 msec

Auditory tone indicates
line to be remembered
(e.g. top line)

Subject repeats cued line

long-term memory so that the information is retained even if rehearsal stops.

Short-term memory involves an interaction between the hippocampus and the areas of the neocortex that are activated by the current sensory input. Activated cortical areas project to so-called "association cells" in the hippocampus that have modifiable NMDA receptors. The constellation of inputs that strongly activate particular association cells have their inputs strengthened. These cells in turn project back to the same areas of neocortex that activated them, so that when areas of the frontal lobe activate the hippocampus association cells, these activate the cortical cells that reproduce a "neural image" of the original input.

Working memory

The original idea of a short-term memory whose contents were loaded from sensory memory has been replaced by a broader concept called *working memory*. Suppose you were asked to multiply 23 by 7. There would be intermediate results while you were doing this calculation that you would retain in a rehearsed short-term memory, but whose origin was not sensory memory but the result of a mental calculation. This broader idea of a working memory is a short-term memory whose contents can come from sensory input or mental thoughts.

In this working memory scheme, the central executive controls the contents of short-term memory via some representation in the dorsolateral prefrontal cortex. Information presented visually to be remembered is transferred from the iconic store to working memory via access to the visuospatial sketchpad. The rehearsal process involves the central executive activating the phonological loop, which is the sub-vocal rehearsal process.

Below: *Working Memory.*

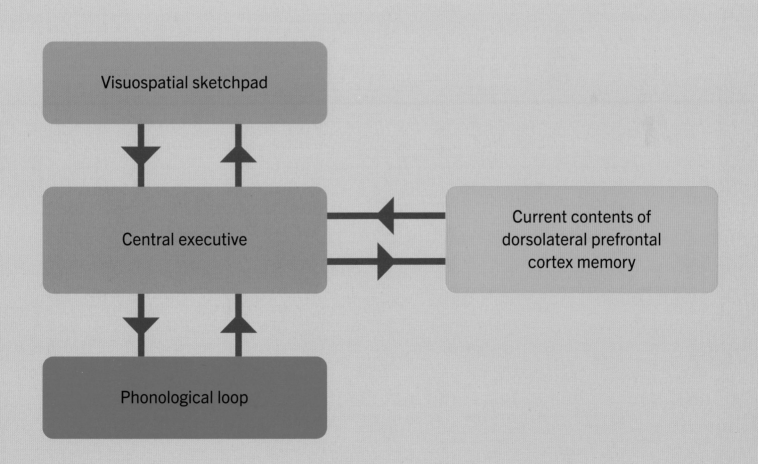

Opposite: *George Miller.*

// Long-term memory

The creation of long-term memory is a process that is gradually strengthened by rehearsal over a period of time from tens of seconds to minutes. The outcome of this process is a memory "trace" that can last a lifetime.

An interesting experiment revealed how sensory decay and long-term memory loading work. Subjects were presented, at one-second intervals, a list of up to ten words. They were then asked to repeat the list in free recall. Subjects were better at recalling items at the beginning of the list (called *primacy*) and at the end of the list (called *recency*). The serial position effect is thus a U-shaped curve when accuracy of recall is plotted against the item's position on the list.

Below: *The serial position effect showing primacy and recency.*

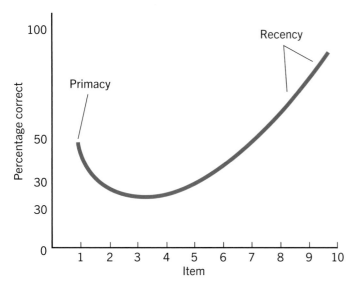

The primacy effect reflects the fact that rehearsal has had more time (rehearsals) to transfer earlier items in the list from short-term memory to long-term memory. This is consistent with the result that when rehearsal is prevented the early items are rapidly forgotten. The recency effect happens because items at the end of the list are available in the sensory-short-term memory buffer having recently been seen, and therefore have not yet decayed.

The ideas about memory presented so far suggest a serial processing system in which sensory inputs move from the sensory store to short-term or working memory, to long-term memory. This idea was formalized in what is called the Atkinson and Shriffrin modal model. The idea is that, of all the things in our visual field sensory register, what we pay attention to gets moved into short-term storage, and then during *rehearsal* the data create a permanent trace in long-term storage.

The importance of the hippocampus

The search for the location of long-term memory, the "engram," was one of the most intense in the history of neuroscience. In the mid-20th century Karl Lashley performed experiments in which he trained rats to learn the path through a maze to find a food reward, removed part of their cortex, and then determined the decrease in maze performance afterward. He did not find that any particular part of the brain held the "engram" for the correct maze path, and concluded that memory was "holistically" represented in some unknown manner throughout the brain.

Below: *The multi-store model of memory.*

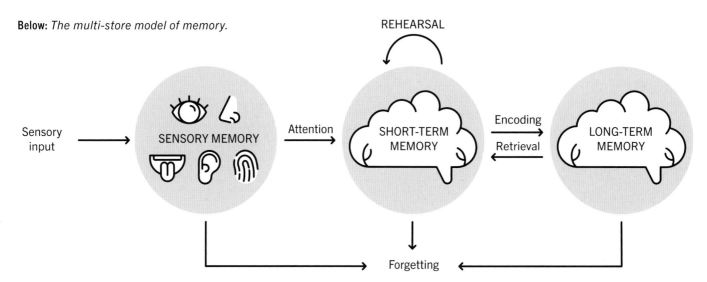

Above: *A rat navigating a maze.*

Lashley was a great psychologist, but not a great biologist, and most neuroscientists today differ with his interpretation of his results. It is now known for example, that rats can use multiple sensory modalities to traverse mazes, including vision, smell, and whisker activation, among others. No single lesion of the brain is likely to destroy all of these different systems. Thus, the failure of the lesions in Lashley's experiments does not imply that memory is not localized at all, but that there are redundant memory systems associated with different sensory modalities in various parts of the brain.

Long-term memory actually appears to be stored in the sensory processing hierarchy of each sensory modality. That is, some of the same sensory neurons that were activated during the learning experience are activated during learning recall. This process depends crucially on the *hippocampus*. One reason we know this is because of an epilepsy patient whose initials are H.M.

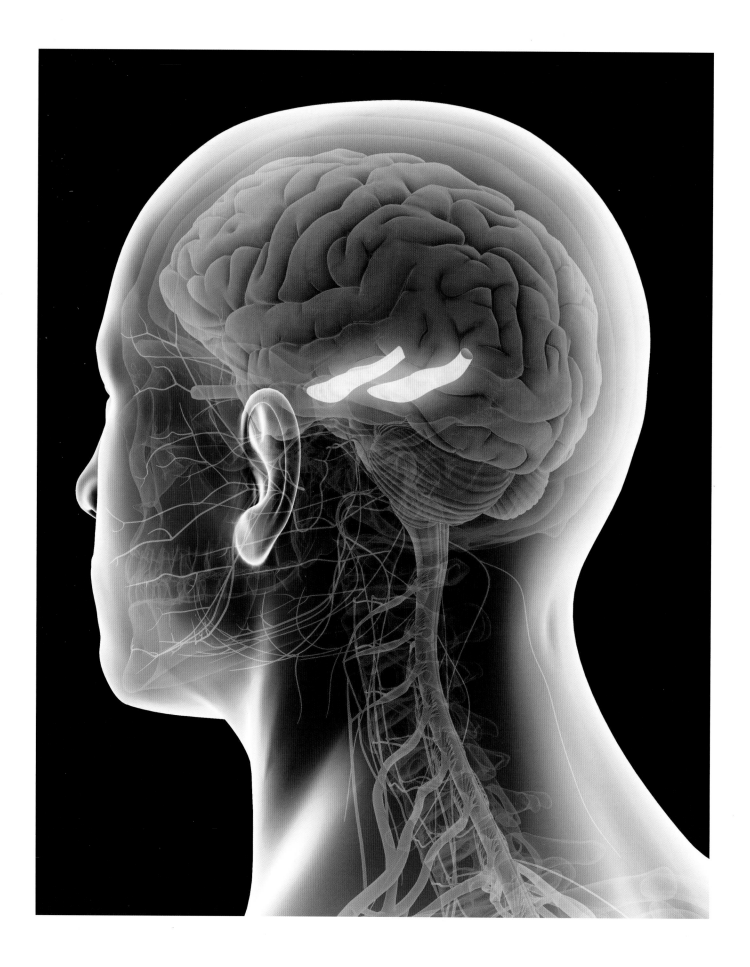

Epilepsy is a debilitating neurological disorder in which self-amplifying neural activity produces uncontrolled neural spiking across the brain. Motor control and consciousness are typically lost in the worst "grand mal" seizures. For reasons that are not currently understood, many seizures originate in the temporal lobe from relatively small brain areas (foci) that initiate the abnormal excitatory runaway activity. If the seizure focus can be identified by electrophysiological recording, surgical removal of the tissue is often successful in reducing seizures.

H.M. was one such epilepsy patient whose temporal lobe-initiated seizures were so frequent and debilitating that normal life was almost impossible. Today, the surgical approach typically involves removing temporal lobe tissue on one side of the brain only, but in H.M.'s case this was done on both sides. Although the surgery was successful in lowering the number and severity of H.M.s seizures, it had a remarkable result that has profoundly changed the modern understanding of the mechanism of memory creation.

H.M.'s long-term memories from his entire life remained intact. This included both general knowledge and autobiographical memory. However, he could form no new long-term memories. As soon as a member of the hospital staff left the room, he had no memory of the person, even if that person saw him every day after the surgery. Memories of anything new that happened more than a few minutes in the past simply did not exist.

The H.M. case, and data since then, show that the process of memory formation involves:

1. The projection of activity from most of the neocortex to association units in the hippocampus.
2. Modification of hippocampus synapses, particularly those involving NMDA receptors.
3. The projection of hippocampal association units back to the areas of neocortex that activated the hippocampal neurons.

Sensory cortical regions project to several structures just

Below: *A brain slice from patient H.M.*

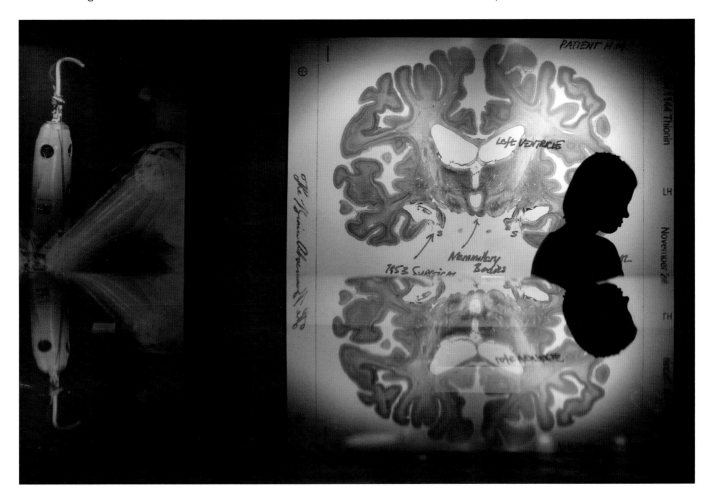

Below: *Hippocampus input and output circuits.*

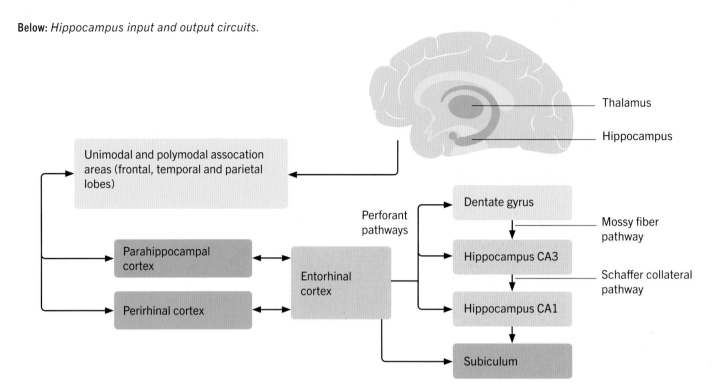

outside the hippocampus, namely the parahippocampal cortex, perirhinal cortex, and entorhinal cortex. These areas seem to reorganize the cortical input for projection into the hippocampus itself. The hippocampus contains association neurons that are activated by combinations of features associated with particular objects, such as their color and shape.

Most of the input stream to the hippocampus contains reciprocal connections, so that those neurons that activated particular neurons in the hippocampus can later be activated by activation of hippocampal neurons. There is also a major output of the hippocampus via a fiber tract called the fornix that terminates in structures called the mammillary bodies that project to the frontal lobe.

The representation of any sensory input, sometimes called the neural image, activates neurons across the cortex for the object's characteristics. In vision, this includes color, shape, motion, and many other characteristics. Neurons from all of these areas project to the hippocampus, in a multi-dimensional matrix. One neuron in the hippocampus gets coincident input from active cortical neurons of the object's color and shape, and will become strongly activated as a green triangle detector. This hippocampal neuron then becomes an "association" neuron for this constellation of properties.

The synapses from the cortical areas to the activated neuron in the hippocampus will be strengthened. Moreover, the projections from cortex to hippocampus are reciprocal, so that the synapses out from these hippocampal association neurons

back to the cortical areas will also be strengthened. After this synaptic modification, activation of those hippocampal neurons (for example, in trying to recall a memory) will activate the same cortical areas activated by the original visual input, reconstructing a neural image or "memory."

Synaptic modification via NMDA receptors

Memory formation based on synaptic modification requires that there exist synapses that can be modified by experience. One such synapse appears to be the NMDA (N-methyl D-aspartate) receptor for glutamate. This receptor has two properties essential for being a modifiable synapse for associative memory:

Below: *Hippocampal association circuits.*

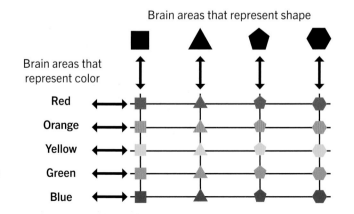

1. Opening the channel requires the coincidence of two inputs.
2. Opening the channel causes long-term modification of synaptic transmission.

The NMDA receptor has a binding site for glutamate that normally would open the channel. However, the charge distribution of the extracellular side of the channel is such that the channel opening weakly binds a magnesium ion (Mg++) that blocks the channel. However, if an adjacent excitatory synapse activates at the same time as the NMDA receptor this depolarization releases the magnesium ion and allows current to flow through the NMDA receptor.

The NMDA channel has a second property that makes it suitable for associative memory formation, which is that it allows both sodium and calcium to enter the cell. The influx of calcium in the postsynaptic neuron brings about long-lasting changes in synaptic efficacy, such as having the NMDA receptor produce a larger excitatory post synaptic potential (EPSP) when the channel is activated.

The NMDA channel is not the only mechanism of synaptic plasticity, but it is one of the most important ones. NMDA receptors are dense in the hippocampus, but occur throughout the neocortex. Another plasticity mechanism is the enhancement of presynaptic release of neurotransmitter resulting from enhanced firing in what is called long-term potentiation (LTP).

Declarative memory

Humans are unique in having not only the non-declarative memory mechanisms we share with other animals, but in having memory reportable by language. This is called declarative or explicit memory. There are two major types of declarative memory: Episodic and semantic.

Episodic memory is typically temporally contextual, that is, it is the memory of specific events or episodes that have occurred in one's life. This type of memory is particularly dependent on the frontal lobe. It is one of the first types of memory to dysfunction in Alzheimer's disease, for example, where a person knows what car keys are, but cannot remember where they put them.

Semantic memory is general world knowledge that is independent of the event(s) during which the knowledge was obtained. Facts like knowledge of colors and shapes by most people, for example, does not include memory of any event associated with when and how that knowledge was obtained. This type of memory is very dependent on the hippocampus, because it is localized throughout sensory and other parts of the neocortex that were activated many times when those world facts were encountered or recalled.

Below: *An NMDA receptor as a coincidence detector.*

Extracellular fluid

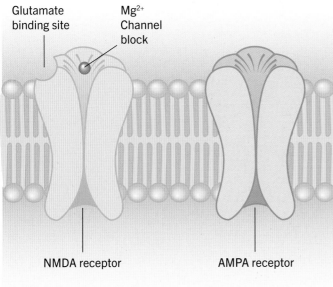

Intracellular fluid

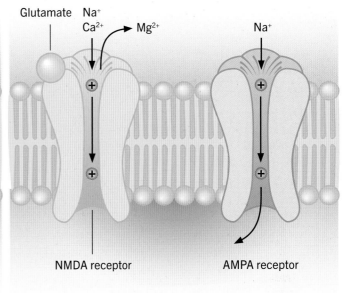

Glutmate and depolarization

// Executive functions

Most executive functions are located in the prefrontal cortex, described as being organized along an axis of abstraction (see pages 120–1), and the frontal lobe, which includes the premotor areas SMA and PMC. An area involved in executive control not in the frontal lobe is the cingulate cortex, which is involved in the control of processing in prefrontal areas.

Prefrontal function

Earlier, we saw how the basal ganglia, acting through the thalamus, chooses an output path of possible actions through the prefrontal and frontal cortex (see pages 121–5). The current goal is represented at the anterior end of the prefrontal cortex. Here, we look at how the action goal itself is chosen. The prefrontal cortex receives inputs from the brain's perceptual and limbic regions that lead to a pathway to motor output. Sensory input to prefrontal cortex arises from parietal, temporal, and high-order occipital (vision) brain areas, brainstem nuclei, and the basal ganglia and cerebrocerebellum. Prefrontal areas are also reciprocally connected to the anterior cingulate cortex.

Below: *Frontal and prefrontal regions of the brain.*

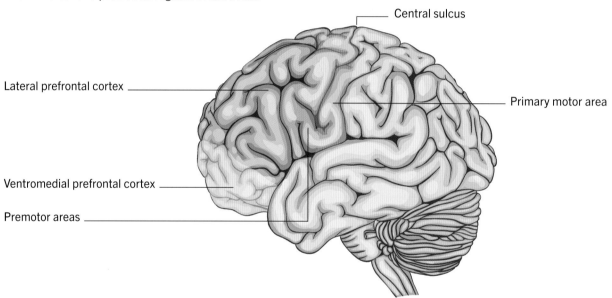

Central sulcus

Lateral prefrontal cortex

Primary motor area

Ventromedial prefrontal cortex

Premotor areas

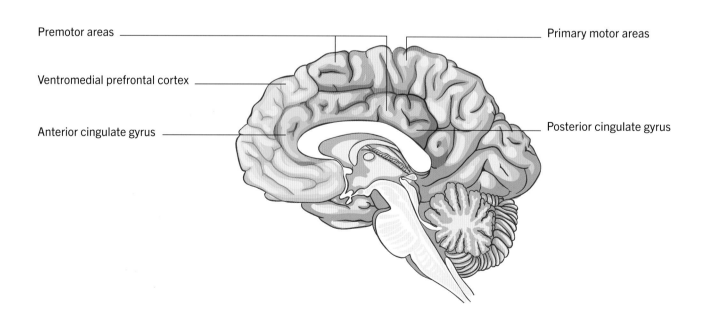

Premotor areas

Primary motor areas

Ventromedial prefrontal cortex

Anterior cingulate gyrus

Posterior cingulate gyrus

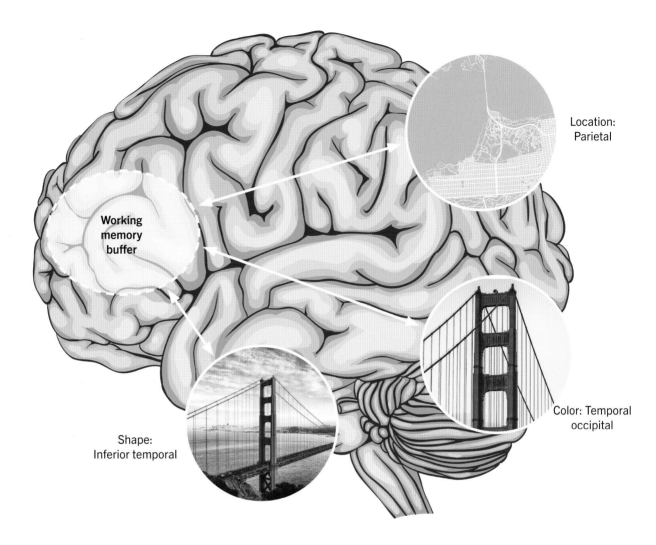

Location:
Parietal

Working
memory
buffer

Color: Temporal
occipital

Shape:
Inferior temporal

Above: *The working memory "image" of the Golden Gate Bridge.*

The net result of all the inputs to prefrontal cortex from external senses of the environment, internal autonomic senses of the organism, and priority from the anterior cingulate is to generate a working memory in the lateral prefrontal cortex, which holds the chosen goal in mind.

Experiments in prefrontal function

A constellation of neurons activated by reciprocal connections from visual areas in parietal, occipital, and temporal lobes to dorsolateral prefrontal cortex might represent an image in the mind (neural image) of the Golden Gate Bridge. The neural image of the Golden Gate Bridge in working memory might be used to accomplish a goal of driving to the bridge, or answering questions about the color and shape of the bridge. This neural image would be used to generate an action hierarchy sequence to meet a goal that involved the bridge.

An experiment by Joaquin Fuster and colleagues recording a monkey's prefrontal neuron during a "cue-delay-go" task illustrates the maintenance of a neural

image in working memory. The monkey receives a cue that indicates a certain response will produce a reward. But the monkey must wait (the delay) until the "go" signal to make the response. Regardless of the length of time, the prefrontal neuron fired during the entire delay (holding the neural image), then ceased firing as the monkey initiated the response, since maintenance of the neural image was no longer needed.

Working memory is more sophisticated than associative memory because it requires the maintenance of an internal image relevant to the goal. In the associative memory task, the monkey learns to choose the food bowl underneath a cross symbol. That is, he learns an association between the cross symbol and the food. After any delay, he simply chooses the bowl beneath the cross symbol. This is simple classical or Pavlovian conditioning (see page 149–150) that even some invertebrates can accomplish.

In the more complex working memory task, the monkey is shown that the bowl on his left contains the food. Then the bowls are hidden during a delay, during which the

Below: *Associative vs working memory tasks.*

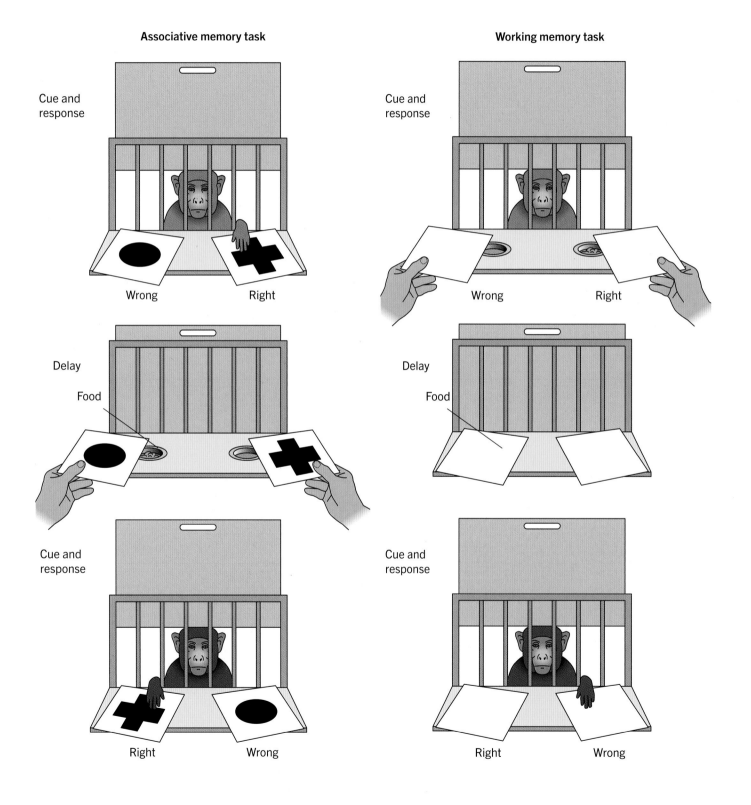

Associative memory task

Cue and response
Wrong Right

Delay
Food

Cue and response
Right Wrong

Working memory task

Cue and response
Wrong Right

Delay
Food

Cue and response
Right Wrong

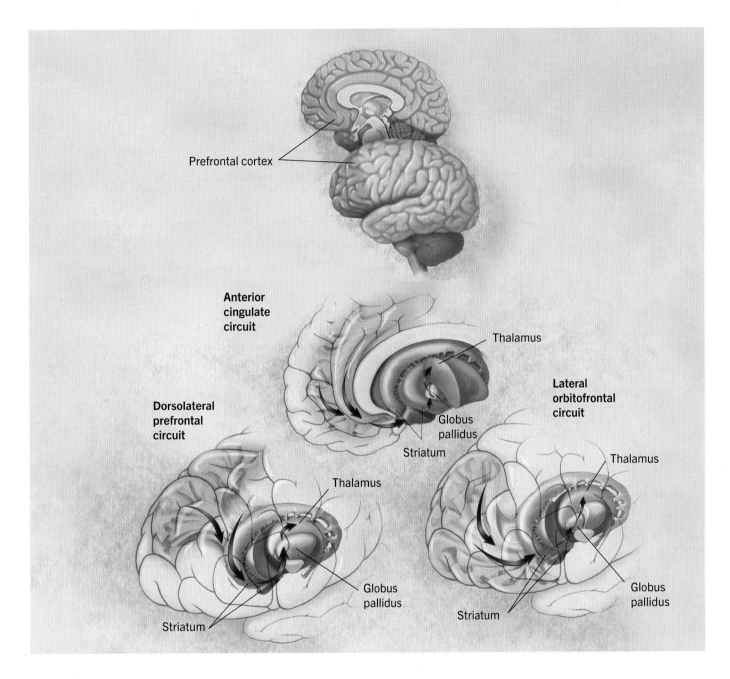

Above: *Prefrontal cortex circuits.*

monkey must remember which bowl had the food, and choose that bowl after being given access. At the choice stage, there is no symbol or information that can be used in an associative manner to pick the correct bowl.

The prefrontal cortex allows goals to be maintained in changing circumstances. This is particularly important when the means being used to achieve a goal must be changed, but not the goal itself.

The Wisconsin Card Sorting test is used to assess prefrontal lobe function. It uses a set of cards that have symbols that vary along several dimensions, such as color,

shape, and number of symbols on the card. In the test, an example card is laid out for each dimension. There is a card with one red triangle, another with two green stars, and so forth. The subject draws a random card from a shuffled deck and is told to guess which example card it goes with, without being told whether the rule is to match color, shape, or the number of symbols.

Normal subjects and those with moderate prefrontal dysfunction usually learn the rule in a few tries. The crux of the experiment is that after 10 correct guesses, the experimenter secretly changes the rule. Normal subjects

Above: *The Wisconsin Card Sorting Test.*

are often initially annoyed, but soon learn the new rule. Subjects with prefrontal damage have extreme difficulty accommodating the new rule, called *perseveration*.

It is frequently difficult to detect a neurological disorder in patients with frontal lobe lesions. They often perform normally on conventional neuropsychological tests of intelligence, and do not display deficiencies in perceptual abilities or speech comprehension or production. But they are prone to perseverative behaviors such as gambling or high risk taking, and often see their job performance and social situation deteriorate due to the inability to maintain important goals versus getting stuck in bad behavior patterns. Some behaviors of frontal lobe patients are like those of teenagers, because the myelination of prefrontal lobe axons is not completed until the mid-twenties. An example is driving too fast because one is in a hurry, despite the fact that the traffic is too dense to drive fast safely, or knowing that there are speed traps on the road.

The anterior cingulate cortex

The anterior cingulate cortex is traditionally considered part of the limbic system (see page 53). The limbic system receives sensory input from the autonomic nervous system as well as from cortical sensory areas. The anterior cingulate cortex also receives input from the memory structures (hippocampus and amygdala), and has reciprocal connections to dorsolateral prefrontal cortex.

The anterior cingulate cortex (ACC) appears to be an executive area that controls the activation of prefrontal areas involved in goal pursuit. Neurons in ACC are activated by:

1. Task difficulty.
2. Pain.
3. Errors in task performance.

A well-known model for this function is the supervisory attention system (SAS) or Norman and Shallice. The SAS is a sub-task selector system that is activated when:

1. Decisions need to be made about carrying out a plan.
2. There are weak links between sensory data and dependent motor behavior.
3. A strong, habitual behavior must be over-ridden by a conscious choice due to incoming data.
4. Errors are being made or the next step is uncertain.
5. Dangerous situations are encountered.

The major selectors of the action path through the frontal lobe are the basal ganglia (see page 55). In the set of actions done in pursuit of a goal there are both unconscious and conscious components. For example, in riding a bicycle to the store, the operation of the bicycle is relatively unconscious. So might be the route selection, out of habit, unless, for example, a street is under construction. The ACC is believed to be involved in the decision to alter the route. A low-level control system involving mostly habitual operations to accomplish a goal is shown as System 1. A higher, more conscious level, shown as System 2, involves the ACC and its processing of task difficulty, uncertainty, and error detection.

Below: *The supervisory attention system.*

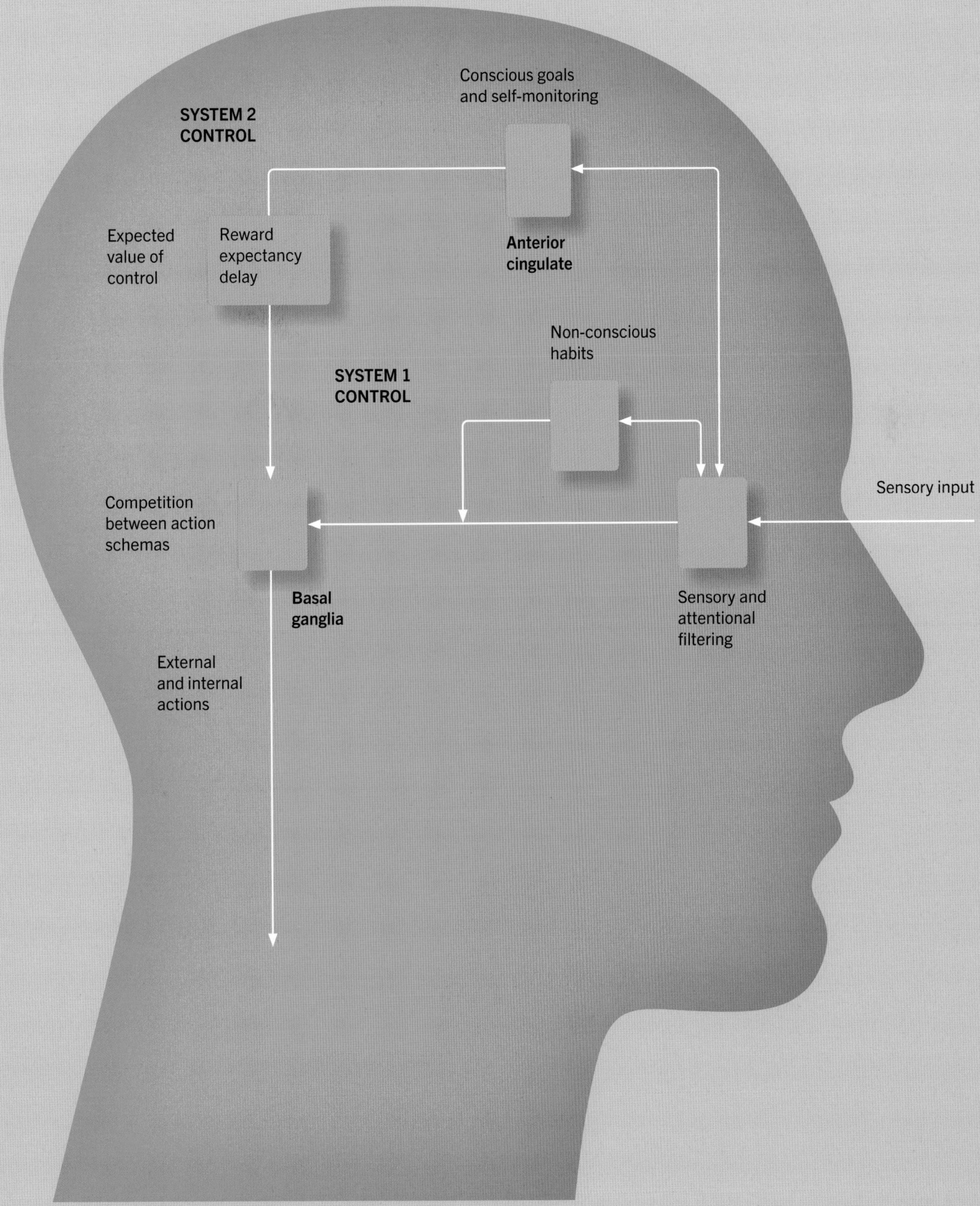

Conscious goals
and self-monitoring

**SYSTEM 2
CONTROL**

Expected
value of
control

Reward
expectancy
delay

**Anterior
cingulate**

Non-conscious
habits

**SYSTEM 1
CONTROL**

Competition
between action
schemas

Sensory input

**Basal
ganglia**

Sensory and
attentional
filtering

External
and internal
actions

// The brain and language

The origin of the species Homo sapiens, and the origin of language, are both typically postulated to have occurred at least 200,000 years ago. One of the most daunting questions in neuroscience and anthropology is what change in the brain from predecessor species permitted language and the rise of technology associated with it?

The brain has two hemisphere that are roughly mirror images of each other. We have already seen how the left side of the brain receives input from, and controls, the right side of the body, and the right side of the brain deals with the left side. We also saw that different lobes of the brain are specialized for different sensory or motor functions with the same right-brain left-body, left-brain right-body organization. The highest cognitive functions of the brain tend to be in anterior areas like the prefrontal cortex. So far, so good.

But, when researchers began learning about the specialization of different parts of the brain for different functions, they discovered another important difference between the two sides of the brain. Language was one of the first high-level functions that could be demonstrated to belong to one side of the brain. The next sections discuss where these differences lie.

Broca's area

The first highly lateralized language area was discovered by the French physician Paul Broca. The area is now named after him. Broca's area is located in the premotor cortex just in front of the primary motor areas that control articulatory mouth parts such as the tongue, used in speech production. Although there is a representation of these areas on the right side of the brain, Broca discovered that a patient, Lazare Lelong, who had suffered damage to the left side of his brain, could not speak properly. His voluntary utterances were limited to a single word-like phrase, "tan." Right-handed patients with damage to the corresponding area on the right side show only minimal speech deficits. Thus, speech

Below: *Broca's (blue) and Wernicke's (green) areas.*

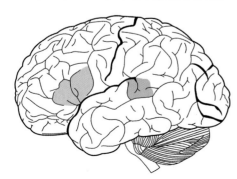

production depends crucially on a specific left-side frontal area in premotor cortex. This is true for virtually all right-handed people and the majority of left-handers.

The term aphasia describes disorders in the ability to communicate. Damage to Broca's area (Broca's aphasia) primarily involves difficulty in language expression.

Howard Gardner gave an example of a patient suffering from Broca's aphasia in his book *The Shattered Mind* (1974). He asked "Were you in the Coast Guard?" The patient replied "No, er, yes, yes … ship … Massachu … chusetts … Coastguard … years", before holding up his hand twice to indicate 19. It is clear from what he does say that he understood the question, and remembers facts about being in the Coast Guard, but is unable to put together coherent, grammatical sentences about what he remembers.

Wernicke's area

A second area, associated with language comprehension, is located at the border between the temporal, parietal, and occipital lobes, and is called Wernicke's area. Patients with damage to this area have difficulty understanding speech, and their own speech is what is called a "word salad" of nonsensical sentences.

When a patient suffers from Wernicke's aphasia, the patient gives a superficially fluent answer to the clinician's question, but the sentences make no sense and are full of non-sequiturs.

Prosody and the right hemisphere

Prosody is a tone of voice component of spoken speech that does not explicitly exist in written speech. There are two major types of prosody: linguistic and emotional. Linguistic prosody has to do with accent and emphasis of syllables within words, and words within sentences. Emotional prosody conveys information beyond the words of the sentence about things like the speaker's belief in what is being stated. For example, if you were to ask most adults if they believed in Santa Claus, many might reply "right!" with a prosody that indicates the opposite of what the answer literally indicates.

Prosody processing depends on right hemisphere brain areas homologous to Broca's and Wernicke's areas of the left brain. This indicates a dual pathway processing of language, with the left hemisphere parsing grammar and word meanings, and the right hemisphere picking up tone of voice and other prosodic information that also conveys speaker information.

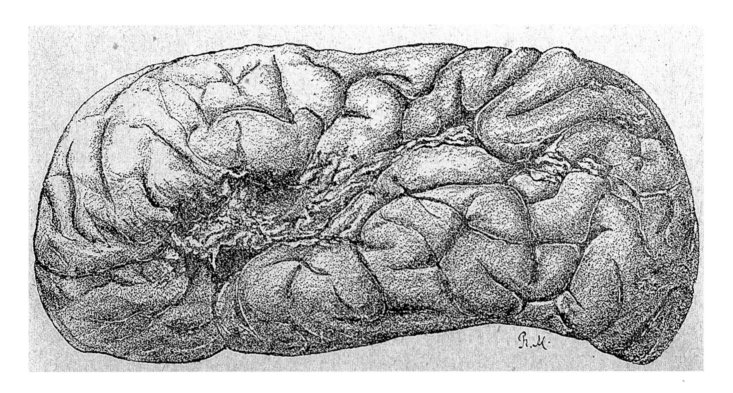

Above: *An illustration of Tan's brain.*

Below: *Broca's and Wernicke's areas are connected by a collection of nerves known as the* arcuate fasciculus.

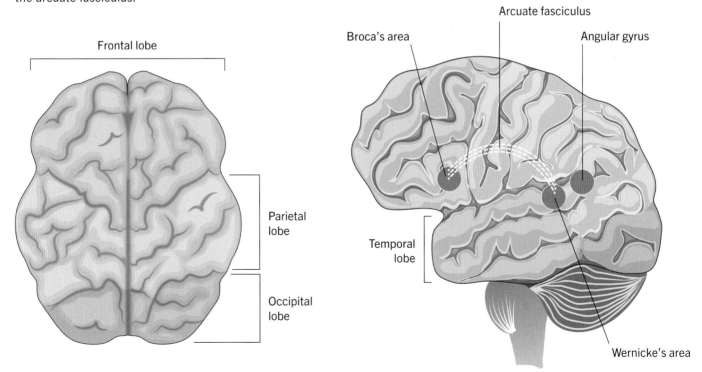

// Brain lateralization

There is no large-scale significant structural difference between the left-side language-enabling half of the brain and the right side. Neither are there any special neural types or grossly different neural circuits on the two sides. Moreover, about 40 percent of left-handers have most language function located on the right side of the brain, and many people who suffer significant left side brain damage early in infancy develop normal language function using the right side of the brain.

It appears, therefore, that whatever brain circuits in the left side of the brain enable language are subtly different from those on the right side. This also seems to be the case for spatial skills, that are much better handled by most people's right than left brain sides. One approach to trying to understand right versus left brain side abilities is to also study how the right side of the brain appears to be specialized for spatial tasks.

Studies of patients with damage to the right side of the brain, particularly in the right parietal lobe, revealed a left-right asymmetry in which the right side of the brain was crucial. This concerns the ability to notice and attend to objects in the left visual field. The inability to deal properly with the left side of space is called a neglect syndrome.

Experiments in unilateral brain damage

In the line bisection test, the patient is given a pencil and a piece of paper with a number of short horizontal lines on it and told to draw a short vertical line through the middle of each horizontal one. The patient is free to move his or her eyes throughout the procedure.

The line bisection test yields two remarkable results:

1. The patient completely misses several lines on the left side of the paper.
2. The patient systematically draws the bisector line too far to the right, as though they don't see some of the line on the left side.
These results typically occur for right, but not left side parietal lobe damage.

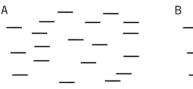 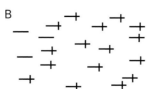

Test paper
(with horizontal lines on it)

Patient bisections
(Vertical lines)

Right side parietal lesions interfere not only with the ability to notice things in the left visual field, but even to imagine them. In a famous experiment by the Italian neuroscientist Edoardo Bisiach and colleagues, a patient with damage to the right parietal lobe was asked to imagine standing at one end of an Italian piazza with which he was very familiar, and describe all the buildings he could see from that vantage point. The patient described only buildings that would be in the right visual field (which would have been projected to the left, undamaged side of the brain). He was then asked to imagine standing at the opposite end of the piazza. He then described only the complementary set of buildings that from this point would have been in his right visual field.

Temporary disability of the brain

The previous examples of brain lateralization came from patients with unilateral brain damage, that is, damage restricted to one side of the brain. Similar results about lateralization have been found when one side of the brain is temporarily disabled, such as with an anesthetic.

Prior to neurosurgery, it is crucial to determine whether language is located in the left or right hemisphere of a particular patient, because a minority of the population, especially left-handers, have language on the right side. A test called the Wada test can determine this. The left and right common carotid arteries supply blood to the left and right sides of the brain almost exclusively.

If an anesthetic such as sodium amytal is injected into the left common carotid artery, and language in the patient is on the left side of the brain, the patient cannot verbally respond to questions, such as whether a spoon has been placed in his left hand. Even after the anesthetic has worn off, the patient cannot verbally identify what was placed in his left hand during the time the left hemisphere was anesthetized. However, at this time, the patient will reach for the correct object with his left hand (controlled by the right hemisphere), indicating that the right hemisphere "remembered" the spoon and could execute a behavior based on this memory.

Split-brain insights into brain lateralization

Results from split-brain patients yield similar results, and some new insights into lateralization. As we saw earlier (pages 136–7), when words or images flashed in the left visual field (that project to the right side of the brain) they did not reach verbally reportable conscious awareness, although

Left: *The results of a standard neurological test for right parietal lobe damage called the line bisection test.*

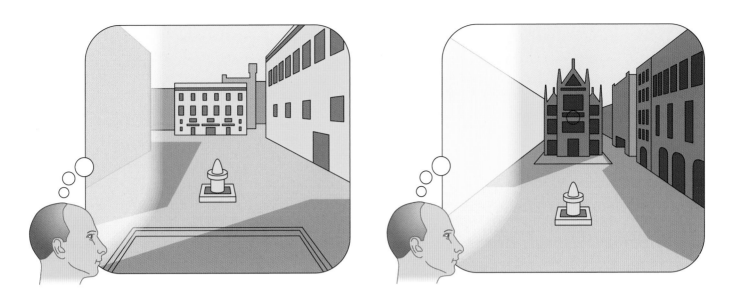

Above: *The Bisiach visual experiment: when imagining a familiar location, the patient could only describe the buildings that appeared in his right visual field.*

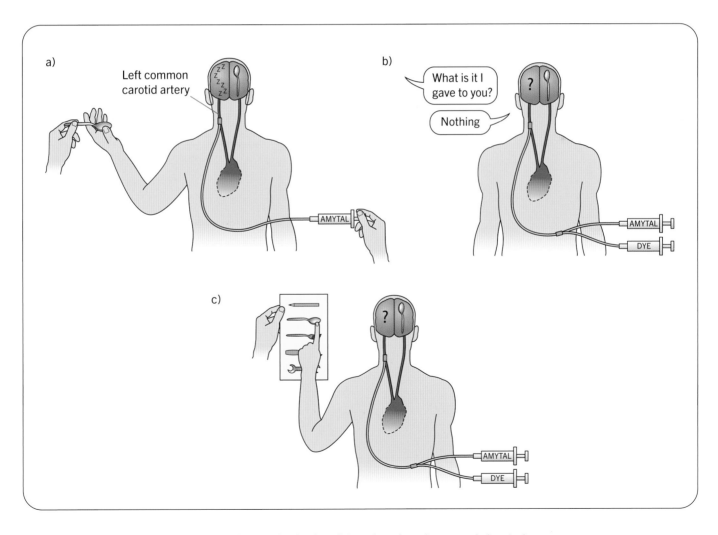

Above: *In the Wada test, one hemisphere is anesthetized to determine where language is located.*

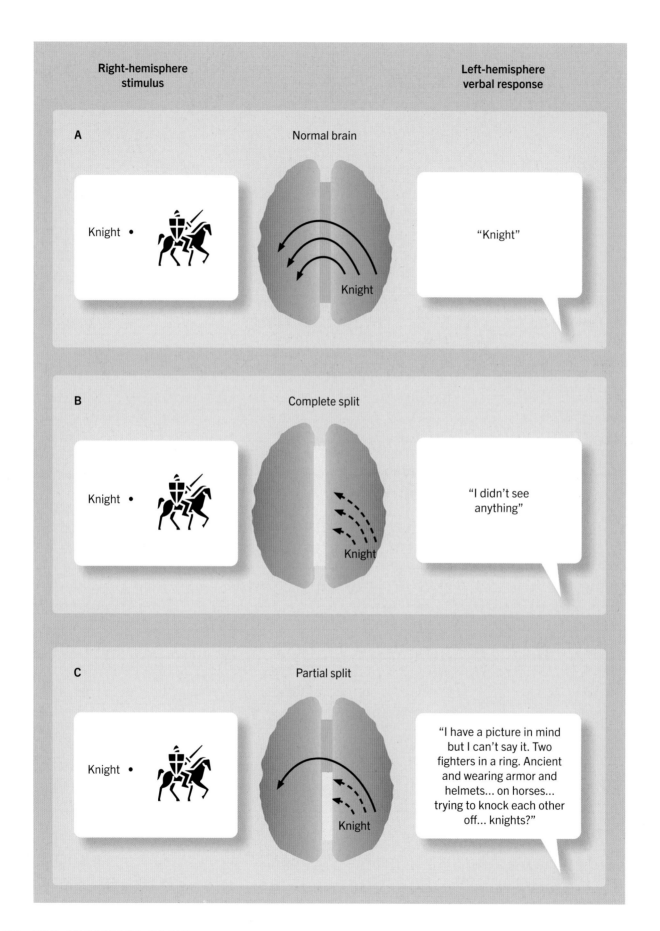

the left arm, controlled by the right side of the brain, could properly choose an object based on the word of which the patient is not conscious.

An interesting case occurs when a partial section is made of the corpus callosum, severing low-level visual interhemispheric connections, but sparing high level ones. Normally if a word is flashed in the left visual field that projects to the right side of the brain, the right brain side sends the information across the corpus callosum to the left side, which can make verbal responses about the identity of the word.

In a complete split of the corpus callosum, no information crosses the corpus callosum from right to left, and the patient can make no verbal report about the word presented in the left visual field. The interesting case is the partial split. Low-level visual processing, such as done by the occipital lobe area V1, sends fibers across the posterior portion of the corpus callosum, while higher, more abstract processing fibers cross more anteriorly. When a patient with a posterior only corpus callosum section was shown the word "Knight" in the left visual field, he had no conscious awareness of the word identity. However, high-level processing of the word on the right side that evoked associate concepts of the word knight were activated, and these crossed the spared anterior portion of the corpus callosum. The result was that the patient began verbally discussing concepts related to the word, and eventually, from his own discussion, guessed that the word presented was "Knight."

Thus, normally, the left side of the brain receives information about low- and high-level attributes of a sensory stimulus. When receiving only high-level information, concepts are formed and verbally accessible without access to the low-level stimulus identity.

Studies of split-brain patients have shown a consistent superiority of the right hemisphere (in left hemisphere language-dominant people) for certain spatial skills. For example, one can ask a split-brain patient to arrange a set of blocks to match the pattern on a picture. Split-brain patients normally do this quite easily with their left hand, controlled by the right hemisphere, but typically struggle at the same task when required to do it with their right hands (controlled by the left hemisphere).

Theories of lateralization

There are a number of theories about lateralization. Specifically, researchers have sought to determine whether the large differences in the two hemisphere's abilities in language versus spatial tasks is caused by some underlying

specialization difference that has implications beyond these two specific kinds of tasks. The hope is that this might tell us about the neural basis for left-language, right-spatial specialization.

Holistic versus detail processing

Prosodic information is processed by the homologous areas of the right hemisphere to those in the left hemisphere that process or produce language. This finding has led to various hypotheses about left versus right processing "styles" or strategies. Most of these proposed schemes suggest that the two hemispheres generally work in parallel by dividing the processing such that the left hemisphere takes care of details and rules, and the right hemisphere looks at the overall picture of what is going on in the input. The corollary of these hypotheses is that, although most tasks require processing by both sides of the brain, some tasks require much more processing by either the left or right brain, and are therefore much more compromised by damage to one side or the other.

Local versus global special processing

Spatial image processing is thought to be right brain dominated. But even within this domain are many scales of detail, from the overall appearance of the image to constituent parts of the image. David Navon (and later Lynn Robertson) and colleagues created images in which a "global" image, such as a triangle, was composed of small components of a different shape, such as little squares.

Patients with either right or left hemisphere damage were asked to reproduce the figure. Patients with right hemisphere damage (but whose left hemisphere was undamaged) produced figures that preserved the identity of the small components, but completely missed the overall shape. Patients with left hemisphere damage drew the overall shape of the triangle, but completely missed that it was composed of little square components.

Below: *The spatial scale perception experiment.*

Non-linguistic	Intact left brain	Intact right brain
Target stimulus	Right-hemisphere patients	Left-hemisphere patients
A	B	C

Prototypes versus exemplars

A variant of the left-right processing styles theory comes from classification experiments. If someone is given an image to classify, such as an animal like a dog, there are two major ways a decision can be made about the image:

1. The image can be compared to some *prototype* image in the mind that has a kind of average set of features that constitute what is a dog that can be described verbally.
2. The image can be compared to memories of all known specific dogs, from dachshunds to greyhounds, called *exemplars*.

In prototype versus exemplar theory the left side of the brain tends to use prototypes, whereas the right side uses exemplars. Prototypes are describable verbally: dogs have four legs, a head, a moderately long tail, and so forth. An example where prototype matching does not work well is if someone is asked what a penguin is. The prototype most people have for birds is that they can fly. Penguins violate this prototype, but we have exemplar knowledge of this specific bird as one that does not fly. Exemplar knowledge and cognition are postulated to be used by the right side of the brain.

Below: *Prototype versus exemplar matching.*

Papillon

Smooth dachshund

Wire fox terrier

Long nose

Perky ears and high energy

Soft coat

Prototypical dog

Facial hair and big pink tounge

Coat clour

General body proportions

Old English Sheepdog

Boxer

Chesapeake Bay retriever

Matching versus maximizing

Prototypes are essentially abstractions about a set of data that follow describable rules. Another situation that illustrates the left hemisphere's tendency to construct and follow rules is the matching versus maximizing experiment. In this experiment subjects were asked to guess the color of an upcoming stimulus after observing several stimuli in a random sequence. In this sequence red stimuli appeared 75 percent of the time, green 25 percent. There are two main strategies for this task:

1. Frequency *matching*: guessing red 75 percent of the time and green 25 percent.
2. *Maximizing*: always guessing red.

The maximizing strategy is the best because the guess will be correct 75 percent of the time. This is what animals do, which has even been demonstrated for fish. But humans use the matching strategy, and thus do worse on this task than fish! Humans appear to be compelled to postulate some pattern or rule in the event sequence and then respond according to this rule. When this experiment was done on split-brain patients (Wofford) it was found that the right hemisphere used the correct maximizing strategy, but the left hemisphere used matching.

Categorical versus coordinate descriptions

A similar dichotomy for image analysis and perception has been referred to as *categorical* versus *coordinate*. If presented with the "Target" image seen below, the categorical information is the verbal description that the small cat is to the left of the large cat. The coordinate description, somewhat harder to verbalize, is about the relative distance between the two cats.

If, after seeing the Target image, a person is shown the upper "categorical transformation probe," it is hypothesized that the left brain brings to consciousness the rule violation that the small cat was to the left of the big cat.

The "coordinate transformation probe" in the lower right does not violate the rule, but the distance has been distorted compared to the target. This violation is thought to be processed by right brain activity.

Below: *Categorical versus coordinate representations.*

Anecdotally, the difference between categorical and coordinate representations often appears when one asks for driving directions. A person giving left-brain categorical directions will say something like, "go to the second traffic light, turn left, and then turn right at the third stop sign." A person giving right-brain coordinate directions might say, "Go about a mile and a half and turn left. Go three blocks and then turn right."

Differences between the Sexes

Much has been written in popular treatments of brain lateralization about sex differences. Data exist that suggest that males tend to be more right-brained (holistic), and women more left-brained (verbal and linear thinking). A few points are in order about this idea.

One fact that is not well understood in many discussions about aptitude and population statistics is the effect of variance in the population on the distribution of extreme aptitudes. Two populations having exactly the same mean but different variances are plotted. Even though the two populations have exactly the same average aptitude, almost all of the very low and very high aptitude individuals are from the high variance population.

Polygamy

Some data, and arguments from Darwinian selection mechanisms, suggest that variance in many aptitudes tends to be larger in males than females. Most mammals are polygamous, with a few males contributing disproportionally to the next generation of offspring, whereas virtually all females that reach adulthood reproduce. It is also the case that most genome mutations are deleterious, not beneficial. These two facts suggest that there should be strong selection pressure across the female population to minimize or suppress mutations because of their net negative overall result. For males, however, those with the small percentage of beneficial mutations will produce many more offspring, so the lack of fitness in many males is of little evolutionary consequence.

There are two closely related species of voles (a type of wild rodent), called meadow voles and prairie voles. Meadow voles are polygamous, whereas prairie voles are not. The females of both vole species occupy nest territories of about the same size, but, since the meadow voles are polygamous,

Below: *The graph shows two populations with the same mean aptitude but very different variances.*

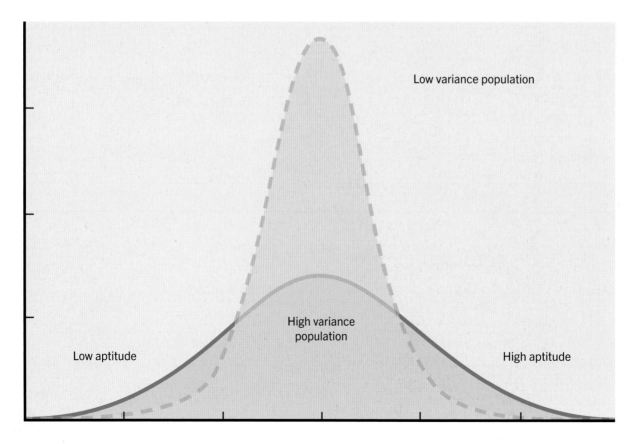

// **176** THE CONSCIOUS BRAIN

their ranges that include several female territories are much larger than their prairie vole cousins. Experiments have shown that the male meadow voles are better at navigating mazes than the male prairie voles, but there is no difference between the two female species. The size of the hippocampus is also much larger in male meadow voles than male prairie voles and female voles of either species. Thus spatial navigation, a right brain coordinate representation, is superior in the polygamous male versus female meadow voles.

Sex hormones and brain development

Sex hormones are also a major mechanism of differential brain development. Testosterone enhances right brain development, while estrogen promotes left. For example, females in male-female fraternal twin pairs exposed to testosterone in utero from their twin brother exhibit long-term brain changes and behavior differences compared to other females.

Above: *A meadow vole.*

Lateralized specialization

There is also evidence from neurological case studies that males generally exhibit more lateralized specialization than females. Strokes and other forms of unilateral brain damage generally produce worse results in males than females, and poorer recoveries. Functional specialization in lateralization has benefits and costs. The benefit is greater aptitude (depth) of processing in the specialized domain. The cost is greater susceptibility to injury or developmental problems. Therefore people with high lateral specialization, mostly males, tend to excel in dealing with problems by depth of analysis, whereas people with less lateralization, mostly females, tend to solve problems more by a breadth approach. These are, of course, just statistical tendencies that oversimplify all of the complex mental activity involved in problem solving that include complex interactions between nature and nurture.

// Neural dysfunctions, mental illness, and drugs that affect the brain

Many things can go wrong with the brain, the most complicated structure we know of in the universe. Disorders of cognition, emotion, or learning are generally called mental illness. Notable types of mental illness include depression, schizophrenia, and bipolar and obsessive-compulsive disorders. Some mental disorders are clearly based on underlying neurological dysfunction, which can be genetic, environmental, or some combination of these. Traumatic events or abuse can also trigger dysfunction in an otherwise normal brain. If one includes substance abuse in the spectrum of mental illnesses, mental illness may be the single costliest health problem in the U.S. and the world.

There are many cases where there appears to be a genetic susceptibility that does not lead to a mental disorder unless triggered by something in the environment. Schizophrenia is one example of the latter.

Mental disorders linked to genetics

There are a number of mental disorders known to be based on genetic abnormalities. Many of these are called syndromes or developmental disorders. Most are associated with reduced intelligence, as well as other anatomical and physiological dysfunctions outside the brain.

Autism

Autism has multiple genetic causes that range in severity and specific characteristics in different individuals. It is thus now called a spectrum disorder. It occurs about four times more often in males than females. Male incidence is greater than 1 in 100, and appears to be increasing for unknown reasons.

Characteristics of those on the autistic spectrum are consistent with weak left brain function, but in some cases, superior right brain function. Weak left brain functioning of those exhibiting autistic spectrum disorder results in poor social skills and lack of empathy. Enhanced right brain skills include the superior ability to recognize (holistic) patterns and to broadly grasp complex technical systems. Some argue that enhanced right versus left brain function is less damaging to females than males because males start out farther along that continuum to begin with.

At the mild end of the spectrum, symptoms usually involve mild social ineptitude and somewhat poor language skills but sometimes above average intelligence in right brain dominant tasks such as in technical or artistic areas. Many so-called "savants" who have extraordinary memorization, calculating,

Above: *Leo Kanner, who discovered autism.*

or artistic skills are autistic. One constellation of mild symptoms is called Asperger's syndrome, which is an autism type without significant language dysfunction.

Severe forms of autism are debilitating. Sufferers have extremely poor language function, almost no ability to engage in social interactions, and often constantly engage in repetitive behaviors such as continuous rocking back and forth. There is no cure for autism, but therapies aimed at increasing social interaction can sometimes improve social function.

Fragile X syndrome

This disorder stems from an unusually frequently occurring mutation on the X chromosome (leading to the name births, with the incidence being greater in older parents who often accumulate more chromosome damage during aging. People with Down syndrome typically show moderate to severe mental retardation and reduced life expectancies due to the mutation affecting non-brain organ systems. Those with Down syndrome who live as long as 50 years have a high incidence of early Alzheimer's disease.

Williams syndrome

Williams syndrome is a genetic neuro-developmental disorder characterized by mental retardation, except in language skills. It is caused by a deletion of a number of genes on chromosome 7. Individuals with Williams syndrome are highly verbal and excessively sociable. Facial physical traits associated with this syndrome include an elfin facial appearance with a low nasal bridge.

Above: *Hans Asperger.*

"Fragile X.") It is the most common cause of inherited intellectual disabilities, including one type of autism. It typically manifests in reduced prefrontal function and some degree of mental retardation. Fragile X is associated with several distinct physical, emotional, and behavioral traits. Physical facial features include elongated faces, prominent jaws and large protruding ears. Fragile X is also associated with flat feet and poor muscle tone. Behaviorally, fragile X individuals exhibit social anxiety and gaze aversion.

Down syndrome

This is caused by the presence of part or all of an extra chromosome 21. Its incidence is about 1 per 1,000

Below: *Fragile X facial features.*

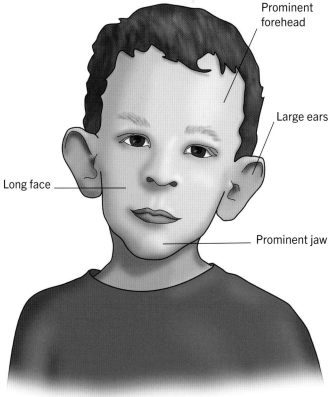

Prominent forehead

Large ears

Long face

Prominent jaw

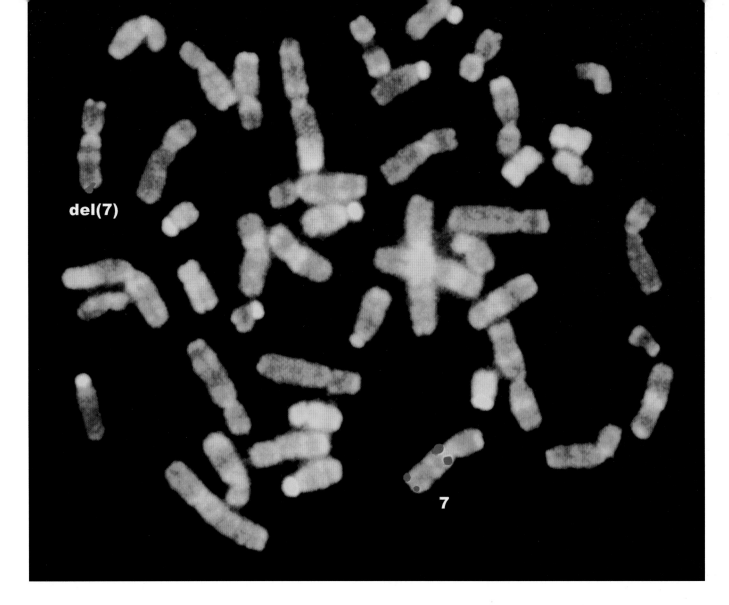

Above: *Williams syndrome chromosomes.*

Rett syndrome

Rett syndrome is a developmental brain disorder that primarily affects females (affected males usually die in utero). During development there are reduced levels of the neurotransmitters dopamine and norepinephrine, and a progressive loss of language and motor skills. Physical features of the syndrome include small hands and feet and a tendency toward microcephaly. Behavior traits include repetitive hand wringing, poor verbal skills, and a tendency to have scoliosis.

Schizophrenia

The term schizophrenia was once applied to what is now called "multiple personality disorder." The term now refers to mental disorders in which a person's thoughts depart seriously from reality. This departure from reality often involves delusions and hallucinations, such as hearing voices in one's head. Overt schizophrenic symptoms typically begin in early adulthood, often with no prior evidence of any mental abnormality. This onset time is when myelination in the frontal lobe is being completed, suggesting an organic, genetic underlying cause.

Although some symptoms of schizophrenia are mitigated by drugs, the disorder is not curable and symptom mitigation requires lifelong drug treatment. Severe schizophrenia often requires hospitalization.

Symptoms of schizophrenia fall into two general categories:

1. *Positive symptoms* are the delusions and hallucinations. Voices heard inside the head sometimes command sufferers to commit inappropriate acts and are associated with delusional beliefs. fMRI recordings have shown that a schizophrenic's auditory cortex is active during the hearing of internal voices, suggesting why it is difficult for them to distinguish hallucinations from reality.
2. *Negative symptoms* refer to withdrawal from and failure to engage in social interaction. They are associated with flat affect, social withdrawal, and loss of motivation, including anhedonia (the inability to experience pleasure). The constellation of symptoms also includes failure to attend to hygiene and many routine life maintenance activities.

Below: *Possible symptoms of schizophrenia.*

| Delusions of grandeur | Hallucinations | Thought broadcasting | Apathy |

| Hearing voices | Catatonia | Disorganised thinking | Persecutory delusions |

Schizophrenia runs in families and clearly has a genetic component. This genetic component may require some unknown environmental trigger, however. If one identical twin has schizophrenia, the other has about a 50 percent chance of developing it. A purely genetic disease would show nearly a 100 percent concordance rate in identical twins. No one has any good idea what this trigger might be, or whether the trigger occurs at the onset of the disease, or much earlier in life before symptoms appear.

Pharmacological treatments (clozapine, quetiapine, risperidone, and perphenazine) for schizophrenia are somewhat successful in the alleviation of positive, but not negative symptoms. Many of the drugs that mitigate schizophrenic positive symptoms increase acetylcholine levels in the brain. Interestingly, the nicotine elevation produced by cigarette smoking also acts as an agonist for acetylcholine. Smoking is prevalent among schizophrenics, perhaps as a form of self-medication. Unfortunately,

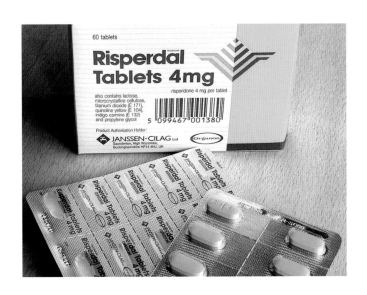

Above: *Risperidone, a drug that can help alleviate the symptoms of schizophrenia..*

the efficacy of drug treatments is highly variable across the schizophrenic population. In addition, many of the schizophrenic drugs not only do not work well in many schizophrenics, but also have serious side effects like weight gain and cardiovascular problems.

Obsessive compulsive disorder

OCD is an anxiety dysfunction. Intrusive thoughts, such as fear of germs everywhere, lead to repetitive behaviors such as constant hand washing. Sufferers of extreme OCD can become so paranoid that they are psychotic. The overall incidence in the population is about 2 percent. Some clinicians believe that OCD, like depression, is linked to low serotonin levels, and some OCD sufferers do have genetic mutations that cause low serotonin levels. Thus OCD is also treated with drugs that elevate serotonin levels by inhibiting its reuptake from synapses, known as SSRIs (selective serotonin reuptake inhibitors).

Environmentally caused mental disorders

Mental disorders can arise in genetically "normal" brains from various types of damage during development or later. Damage can be physical, but also mentally induced, such as from trauma or stress.

Fetal alcohol syndrome

FAS is caused by mothers who drink excessive alcohol during pregnancy. Maternal alcohol crosses the placental barrier and can damage neurons and brain structures in the fetus. In the U.S., evidence of this syndrome is found in about 1 in 1,000 births. Cognitive and functional disabilities associated with this syndrome include attention and memory deficits, impulsive behavior, and stunted overall growth. Exposure to fetal alcohol is associated with a constellation of facial features including a short nose, thin upper lip, and skin folds at the corner of the eyes.

Maternal stress and epigenetics

The children born to chronically stressed mothers tend to have more cognitive and emotional problems. These include anxiety, hyperactivity, attention deficits, and delay in language acquisition. The mechanism for these effects may be elevated levels of the stress hormone cortisol in the mother, which has detrimental effects on the developing fetal nervous system.

Recent research has suggested that maternal stress may have epigenetic effects on the fetus. Virtually all cells in any organism have the same DNA, but different cell types are so because only a subset of the total DNA is expressed for that cell type. Normally the subset of genes to be expressed in a final cell type is fixed during development by what is called epigenetics, the control of DNA expression in various cells by means of "shielding" parts of the genome from transcription to mRNA by mechanisms such as DNA methylation and histone acetylation.

However, even in mature cells, environment factors can sometimes alter the epigenetic constellation, so that, without changing the underlying DNA content, the subset of expressed DNA is changed. This occurs in some cancers where exogenous chemicals change the epigenetics but not the genetics. It can also occur from stress, possibly by long-term changes in circulating hormone levels. Epigenetic changes in somatic cells can also occur in germ cells, and so become heritable. It is important to note that this inheritance is not Lamarckian in the sense that the environmentally induced change somehow produces a new epigenetic constellation that is better at dealing with that environment. In fact, most of these epigenetic effects, like maternal stress appear to be maladaptive.

Post-traumatic stress disorder (PTSD)

PTSD is an example of a mental disorder caused by a life experience(s) that changes the brain in a maladaptive way. PTSD sufferers have excessive anxiety that develops after psychological traumas. Typically such traumas involve a significant threat to one's physical, psychological, or sexual integrity that overwhelms the ability to cope.

The mechanism by which the events cause the disorder involve an excessive fight-or-flight adrenaline response to any danger that resembles the original threat that persists after the original threat event. This is a kind of sensitization in which any experience even slightly resembling the original danger induces a full-fledged autonomic response with the high cortisol and catecholamine secretion that the original response triggered. Over-activation of the sympathetic autonomic nervous system diverts resources from homeostatic mechanisms such as digestion and immune responses that produces long-term damage to the brain and body. This imbalance can result in chronic stress or insomnia. PTSD is sometimes treated successfully with desensitization therapies, in which the sufferer re-experiences aspects of the original stressor in a controlled environment. Other therapies include anti-anxiety medications that reduce the symptoms of the disorder.

Many mental disorders arise from a combination of genetic susceptibility and some environmental trigger(s). This can exist on a continuum such that people with high genetic susceptibility require only small triggers, while those with low susceptibility require a high trigger to spiral into clinical mental illness.

Above: *Art is sometimes used as therapy for PTSD.*

Depression

Depression is the most serious form of mental illness in terms of total cost, affecting about 15 percent of the population, and costing more than $50 billion yearly in the U.S. Depression appears to be a spectrum disorder varying in severity and in genetic and environmental dependence. All humans are depressed for some time after the death of a loved one, for example, but some recover better and more rapidly than others. It is also the case that depression runs in some families such that it appears to have a hereditable component.

A leading theory of depression is that it results from a deficit in the neurotransmitter serotonin. This is called the monoamine oxidase hypothesis. This has led to the use of SSRI drugs (see page 182) for therapy such as the class of antidepressants like Prozac. Many SSRIs also increase levels of norepinephrine and dopamine. However, this theory has come under increased scrutiny because depressed people do not exhibit abnormally low serotonin levels. Research has also shown that artificially lowering serotonin levels in people without depression doesn't induce it in them.

Many psychotherapeutic approaches to treating depression are based on a downward spiral interpretation model. In this model, either a significantly negative event, or even a marginally negative event received while one is in a state of negative affect, causes a long-term hormonal "setting" to change to a more negative interpretation about the events of life. Each new, even slightly negative event pushes the negative affect further downward, eventually locking in the depressed state. The goal of the psychotherapy is to break the spiral by either insight or behavior change.

Treatments

Not only depression, but also anxiety and substance abuse and eating disorders have been hypothesized to be based a combination of obsessive, "one track mind" thinking and inappropriate behavioral responses to obsessive thoughts. The goals of many therapies are to both reduce inappropriate thoughts, such as feelings of hopelessness, as well as engage in fewer negative behaviors (alcohol use), and more positive ones, such as physical activity.

There has been a resurgence of interest in the last decade of the use of some psychoactive drugs for treatment of psychiatric disorders such as depression, PTSD, and substance abuse. This is a complex issue because substance abuse itself clearly causes psychiatric problems, as evidenced by the epidemic of homeless drug addicts in many large American cities.

Not all psychoactive drugs are the same, however. In particular, psilocybin, and some related drugs, used in

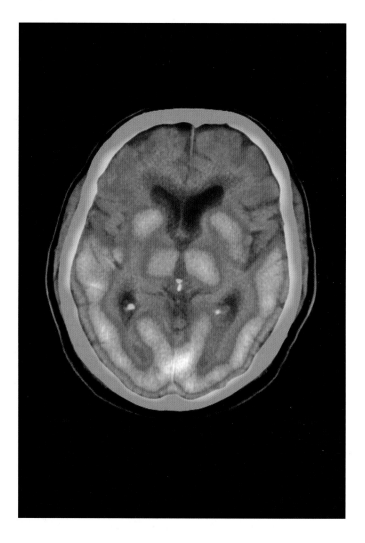

Above: *A CT scan of a brain with depression.*

controlled amounts in clinical settings appear not only not to be addictive, but to be associated with the generation of helpful insights in some people with mental disorders. The positive effects of clinical use of these drugs is obtained in one or two sessions, and, in fact, patients in these studies usually do not want to take the drugs again afterward. This contrasts sharply with substances like cocaine or alcohol that induce dependence in a significant percentage of users. With psilocybin there is the irony of using a drug to mitigate the abuse of other drugs. The jury is still out on this one.

Drug abuse and the brain

Many drugs of abuse increase levels of dopamine in the brain. Dopamine is a neurotransmitter of three major brain systems:

1. The meso-cortical pathway.
2. The meso-limbic pathway.
3. The nigro-striatal pathway.

The meso-cortical and meso-limbic pathways mediate the brain's reward and motivation system through projections to the nucleus accumbens and prefrontal cortex from the ventral tegmental area (VTA). The dopamine released in prefrontal areas is a "reinforcement" neurotransmitter of the brain's reward system for food, sex and otherwise neutral stimuli that become associated with pleasure through learning. Commonly abused drugs such as cocaine, nicotine, and amphetamines increase dopamine in the brain and thus hijack the natural reward system, leading to addiction.

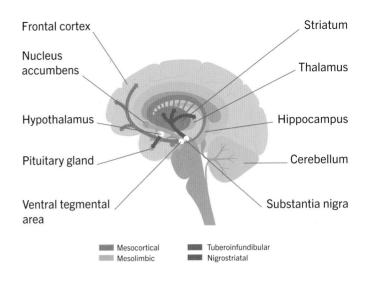

Below: *Drugs increase the level of dopamine in the brain, hijacking its natural reward and reinforcement system.*

Above: *Dopamine pathways in the brain.*

// Glossary

action potential: an electrical pulse that travels down axons (and some dendrites), transmitting signals from a neuron over long distances.

Adrenergic: a type of neuron that uses epinephrine (formerly, adrenaline) as a neurotransmitter. It sometimes is used to refer to neurons that use norepinephrine as well.

Afferents: a nerve axon that sends information inward, toward targets in the central nervous system.

Agonist: A substance that binds to a receptor and activates it.

Amygdala: A nucleus found in front of the hippocampus, involved in organizing emotional responses to events and in storing memories of emotion-laden experience.

Associationism: the theory that the brain is a general-purpose learning machine that stores relationships between concurrent events.

Autoreceptors: receptors for a neurotransmitter released by the same neuron.

Axons: the section of a neuron that carries electrical signals (action potentials) from the cell body to the synaptic terminals.

Behaviourism: the theory that laws of conditioning can explain behaviour without the need to understand brain structure or internal cognition.

Brainstem: the oldest part of the brain between the spinal cord and cerebrum, consisting, from top down, of the midbrain, pons, and medulla.

Catecholamines: neurotransmitters synthesized from tyrosine such as dopamine, epinephrine, and norepinephrine.

Central sulcus: the furrow in the neocortex running in the coronal plane that divides the frontal from the parietal lobe.

Cerebellum: a motor control structure emanating from the pons that is responsible for motor learning and fine movement control.

Cerebral cortex: the large, six-layered area of the top of the mammalian brain divided into two hemispheres.

Cholinergic: refers to neurons that use the neurotransmitter acetylcholine (Ach).

Chordate: any vertebrate that has a dorsal backbone, including species ranging from fish to mammals.

Cingulate gyrus: an area just above the corpus callosum, involved in executive control of other cortical activity and plays a role in monitoring task difficulty, behavioural errors, and pain.

Connexins: proteins that form gap junctions between neurons, usually organized in six subunits that allow ions and small molecules to pass between the two neurons.

Coronal: a section of the brain that divides it into dorsal versus ventral, or anterior versus posterior areas, in contrast to sagittal and horizontal sections. In humans coronal sections are roughly aligned with the face.

Corpus callosum: a massive fibre tract running between the two cerebral hemispheres, consisting of around 200 million axons, that links the two sides of the brain together.

Dendrites: branching sections that emanate from a neuron that receive synaptic input.

Dermatome: the skin area whose receptors arise from the dorsal root ganglion of a single spinal segment.

Endorphins: neurotransmitters, neuromodulators and neurohormones involved in the perception of pain.

Efferents: axons that conduct information away from the brain toward the periphery.

Epigenetics: how changes in environmental influences how your genes work, without changes in the DNA itself.

Excitatory: a neural input that makes the membrane potential toward a more positive electrical value (which promotes the generation of action potentials). This is also called depolarization.

G-protein: one of the guanine nucleotide binding proteins that activate ion channels in cells.

Ganglion: a cluster of neural cell bodies in the peripheral nervous system.

Gastrulation: an early development in embryos in which the single layer of cells known as the blastula transforms into the more complex gastrula.

hemi-channels: the gap junction structure on each cell or neuron that by "mating" with a similar structure on an adjacent cell, forms a pore between the two cells permeable to ions and small molecules.

Hippocampus: a structure in the medial temporal lobe involved in memory, particularly the transfer of memory from short to long-term storage.

Histology: microscopic anatomy of the nervous system (or other body areas), typically using stains to reveal differences between cells or parts of cells.

Homunculus: a neocortical map of the body in either the somatosensory or motor systems.

Indolamines: neurotransmitters with the monoamine structure synthesized from tryptophan, including serotonin, melatonin, and histamine.

Ionotropic: a type of postsynaptic receptor which opens an ion channel when activated.

Inhibitory: a synaptic input that tend to hyperpolarize the postsynaptic neuron or oppose depolarization by an excitatory input(s).

Innervate: to supply some part of the nervous system or body with axons (nerves).

Kinesthesia: the awareness of the position (proprioception) and movement (kinesthesis) of body parts via sensory receptors in the muscles, tendons, and ligaments.

Limbic system: a subcortical system of nuclei and tracts including the hippocampus, amygdala, thalamus and cingulate gyrus involved in modifying instinctual and emotion-driven behaviour.

Localization: the idea that specific functions and behavioural attributes are located in specific areas of the neocortex.

Magnocellular: a type of retinal ganglion cell with a relatively large receptive field and transient responses.

Mechanoreceptors: somatosensory receptors that mediate skin senses such as touch, pressure, and vibration.

Medulla: the lowest of the three divisions of the brainstem, just above the spinal cord.

Mesocortex: a type of cortex phylogenetically older than neocortex that includes the cingulate gyrus and several limbic structures.

Metabotropic: a type of postsynaptic receptor which does not include an ion channel but instead activates a g-protein cascade inside the cell that opens nearby channels.

Midbrain: the highest of the three divisions of the brainstem.

Monoamines: neurotransmitters whose molecular structure involves a single amino group attached to an aromatic ring via a two-carbon chain, such as serotonin and the catecholamines dopamine, epinephrine and norepinephrine.

Muscarinic: a metabotropic type of acetylcholine receptor found in the brain and peripheral nervous system which also responds to muscarine.

Neocortex: see cerebral cortex.

Neuromodulators: transmitters that have modulatory effects and which often affect receptor sites distant from the releasing site.

Neurohormones: transmitters that enter the extracellular space or bloodstream and affect distant targets, often with long lasting rather than rapid effects.

Neurons: specialized cells that, together with glial cells, form the nervous system. Most neurons have dendrites that receive synaptic inputs, and a single axon that carries the output of the neuron to other cells.

Neurotransmitter: a chemical substance released by a neuron to communicate with other neurons by opening ion channels.

Neurulation: the embryonic process in which the neural plate folds up to form the neural tube that later differentiates into the brain and nervous system.

Nicotinic: the major ionotropic receptor type for acetylcholine.

Noradrenergic: neurons that use the neurotransmitter norepinephrine.

Nociceptors: receptors for pain.

Occipital lobe: The rearmost area of the brain, almost all of which is involved in visual processing.

Olfactory: having to do with the sense of smell.

Parietal lobe: the area in front of the occipital lobe and behind the frontal lobe and above the posterior temporal lobe. This lobe processes visual, somatosensory and auditory information.

Parvocellular: a retinal ganglion cell class with small receptive fields and sustained responses, involved in color and fine detail vision, as opposed to magnocellular cells.

Phylum: the taxonomic class below kingdom and above class, including groups like chordates, arthropods, and mollusks.

Phylogenetic: referring to the evolutionary descent of an organism, or a structure within an organism.

Pons: the middle of the three subdivisions of the brainstem.

Primary motor cortex: The area just in front of the central sulcus, involved in driving muscles and controlling voluntary movement.

Proprioception: the perception of joint and body position, mediated by receptors in muscles, tendons, and ligaments.

Rectifying channels: membrane ion channels that allow ion movement in only one direction, either from outside to inside the cell, or vice versa.

Saccades: rapid eye movements made between fixations, typically at a rate of 3-4 per second.

Sagittal: a vertical brain section going from front to back, as opposed to coronal or horizontal sections.

Soma: the cell body of a neuron.

Somatosensory: having to do with sensation at a particular location within the body, particularly in the skin, muscles, and tendons.

Spike: see action potential.

Synapse: the locus of the connection between a presynaptic and postsynaptic neuron or muscle cell. The majority of synapses are chemical, involving release and receipt of a neurotransmitter, but some synapses are electrical, using gap junctions.

Synaptic cleft: the gap between the neuron and its target cell at a synapse, typically about 20 nanometers in size.

Temporal lobe: the lateral lobes of the brain, used for memory, language, and hearing.

Thalamus: a structure that transmits sensory information to the neocortex, motor commands to frontal lobe, and receives information back from these areas. The thalamus and neocortex form a processing system with the thalamus as the controller, and the neocortex as the executor.

Tonotopic map: a map representing where different sound frequencies are perceived.

Vesicles: membrane-bound packets of neurotransmitter whose contents are released at synapses.

Visuospatial sketchpad: in Baddeley's working memory model, it refers to the component which maintains visual images in short term memory and derives information from them.

// Index

Picture Credits

t = top, b = bottom, l = left, r = right

Alamy: 12, 159

Bridgeman Images: 8

David Woodroffe: 115, 137, 164, 166, 171t, 171b

Getty Images: 18, 135, 146, 151

Library of Congress: 17

National Library of Medicine, USA: 178

Princeton: 154

Science Photo Library: 20, 22, 27, 34 (x8), 39, 40, 50, 51, 63, 67, 75, 77, 81l, 84, 99, 100, 106, 110t, 111, 121, 150, 165, 179b, 180, 184

Shutterstock: 14, 23l, 23r, 24, 28, 30, 32, 36, 43, 45t, 45b, 46, 47t, 47b, 52, 53, 55, 56, 57t, 57b, 58b, 59, 60, 61, 62, 65, 68, 69, 70, 71, 72, 76t, 76b, 79, 80, 85, 86, 88, 89, 90, 91, 92, 93, 95, 96t, 97, 102t, 103, 104l, 104r, 108t, 108b, 109, 112, 113, 117, 123b, 126, 127l, 127r, 128t, 128b, 130, 132, 133, 136, 139b, 141, 143t, 143b, 144t, 145, 147, 148, 149, 156b, 157, 158, 162, 163, 169b, 177, 181t, 185t, 185b

Wellcome Collection: 6, 13, 37t, 40, 114, 122, 131, 142, 169t

Wikimedia Commons: 7, 10, 11, 15l, 15r, 19, 21, 34 (x1), 37b, 83r, 88, 98, 140, 144b, 179t, 181b, 183